SOLUTIONS MANUAL
to accompany

MEDICAL INSTRUMENTATION
APPLICATION AND DESIGN

THIRD EDITION

John G. Webster, Editor

Contributing Authors

John W. Clark
Rice University

Michael R. Neuman
Case Western Reserve University

Walter H. Olson
Medtronic, Inc.

Robert E. Peura
Worcester Polytechnic Institute

Frank P. Primiano, Jr.
ECRI, Inc.

Melvin P. Siedband
University of Wisconsin–Madison

John G. Webster
University of Wisconsin–Madison

Lawrence A. Wheeler
Nutritional Computing Concepts

John Wiley & Sons, Inc.
New York • Chichester • Weinheim • Brisbane • Singapore • Toronto

Copyright © 1998 by John Wiley & Sons, Inc.

Excerpts from this work may be reproduced by instructors for distribution on a not-for-profit basis for testing or instructional purposes only to students enrolled in courses for which the textbook has been adopted. *Any other reproduction or translation of this work beyond that permitted by Sections 107 or 108 of the 1976 United States Copyright Act without the permission of the copyright owner is unlawful. Requests for permission or further information should be addressed to the Permissions Department, John Wiley & Sons, Inc., 605 Third Avenue, New York, NY 10158-0012.*

ISBN 0-471-17359-2

Printed in the United States of America

10 9 8 7 6 5 4 3 2 1

Printed and bound by Malloy Lithographing, Inc.

Preface

Solutions were prepared by the contributing authors and compiled by the editor. Page 144 contains a list of 32 suggested laboratory experiments. We welcome your suggestions for corrections and improvements for subsequent editions and printings. In particular, send any lists of homework problems and examination questions that you have found instructive to the editor in the care of John Wiley & Sons, 605 Third Avenue, New York, NY 10058.

John G. Webster

Contents

Chapter 1	Basic Concepts of Medical Instrumentation	1
Chapter 2	Basic Sensors and Principles	8
Chapter 3	Amplifiers and Signal Processing	17
Chapter 4	The Origin of Biopotentials	25
Chapter 5	Biopotential Electrodes	37
Chapter 6	Biopotential Amplifiers	49
Chapter 7	Blood Pressure and Sound	69
Chapter 8	Measurement of Flow and Volume of Blood	79
Chapter 9	Measurements of the Respiratory System	91
Chapter 10	Chemical Biosensors	106
Chapter 11	Clinical Laboratory Instrumentation	108
Chapter 12	Medical Imaging Systems	112
Chapter 13	Therapeutic and Prosthetic Devices	120
Chapter 14	Electrical Safety	136
Suggested Laboratory Experiments		144

Chapter 1
Basic Concepts of Medical Instrumentation

Walter H. Olson

1.1 The following table shows % reading and % full scale for each data point. There is no need to do a least squares fit.

Inputs	0.50	1.50	2.00	5.00	10.00	
Outputs	0.90	3.05	4.00	9.90	20.50	
Ideal Output	1.00	3.00	4.00	10.00	20.00	
Difference	−0.10	+0.05	0.0	−0.10	+0.50	
% Reading	−11.1	1.6 %	0%	−1.0%	+2.4 %	$=\dfrac{\text{Difference}}{\text{Output}} \times 100$
Full Scale	−0.5%	0.25%	0%	−0.50%	+2.5%	$=\dfrac{\text{Difference}}{20} \times 100$

Inspection of these data reveals that all data points are within the "funnel" (Fig. 1.6b) given by the following independent nonlinearity = ±2.4% reading or ±0.5% of full scale, whichever is greater. Signs are not important because a symmetrical result is required. Note that simple % reading = ±11.1% and simple % full scale = 2.5%.

2 Basic Concepts of Medical Instrumentation

1.2 The following table shows calculations using equation (1.8).

Inputs X_i	0.50	1.50	2.00	5.00	10.00	$\bar{X} = 3.8$
Outputs Y_i	0.90	3.05	4.00	9.90	20.50	$\bar{Y} = 7.6$
$X_i - \bar{X}$	−3.3	−2.3	−1.8	1.2	6.2	
$Y_i - \bar{Y}$	−6.7	−4.55	−3.6	2.3	12.9	
$(X_i - \bar{X})(Y_i - \bar{Y})$	22.11	10.465	6.48	2.76	79.98	$\Sigma = 121.795$
$(X_i - \bar{X})^2$	10.89	5.29	3.24	1.44	38.44	$\Sigma = 59.3$ $(59.3)^{1/2} = 7.701$
$(Y_i - \bar{Y})^2$	44.89	20.7	12.96	5.29	166.41	$\Sigma = 250.25$ $(250.25)^{1/2} = 15.82$

$$r = \frac{121.795}{(7.701)(15.82)} = 0.9997$$

1.3 The simple RC high-pass filter:

The first order differential equation is:

$$C\frac{d[x(t) - y(t)]}{dt} = \frac{y(t)}{R}$$

$$\left(CD + \frac{1}{R}\right) y(t) = (CD)\, x(t)$$

$$\frac{y(D)}{x(D)} = \frac{D}{D + \frac{1}{RC}} \qquad \text{Operational transfer function}$$

$$\frac{Y(j\omega)}{X(j\omega)} = \frac{j\omega RC}{j\omega RC + 1} = \frac{\omega RC}{\sqrt{(\omega RC)^2 + 1}} \qquad \angle \phi = \text{Arctan}\, \frac{1}{\omega RC}$$

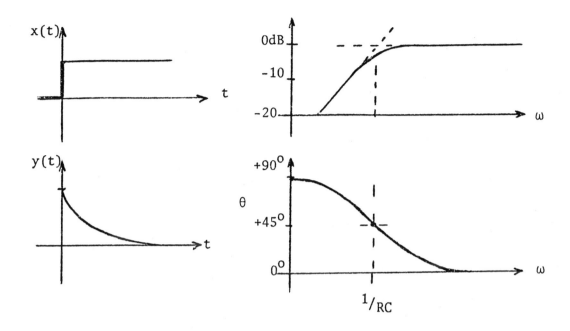

1.4 For sinusoidal wing motion the low-pass sinusoidal transfer function is

$$\frac{Y(j\omega)}{X(j\omega)} = \frac{K}{(j\omega\tau + 1)}$$

For 5% error the magnitude must not drop below 0.95 K or

$$\left|\frac{K}{j\omega\tau + 1}\right| = \frac{K}{\sqrt{\omega^2\tau^2 + 1}} = 0.95\,K$$

Solve for τ with $\omega = 2\pi f = 2\pi(100)$

$$(\omega^2\tau^2 + 1)(.95)^2 = 1$$

$$\tau = \left[\frac{1 - (0.95)^2}{(0.95)^2(2\pi 100)^2}\right]^{1/2} = 0.52 \text{ ms}$$

Phase angle $\phi = \tan^{-1}(-\omega\tau)$ at 50 Hz

$$\phi_{50} = \tan^{-1}(-2\pi \times 50 \times 0.0005) = -9.3°$$

at 100 Hz

$$\phi_{100} = \tan^{-1}(-2\pi \times 100 \times 0.0005) = -18.2°$$

1.5 The static sensitivity will be the increase in volume of the mercury per °C divided by the cross-sectional area of the thin stem

$$K = \frac{\gamma_{Hg} V_b}{A_c} = 2\,\text{mm/°C}$$

where

$$\gamma_{Hg} V_b = 1.82 \times 10^{-4} \frac{\text{cm}^3}{\text{cm}^3\,°C}$$

V_b = unknown volume of the bulb

A_c = cross-sectional area of the column

$A_c = \pi(0.1\,\text{mm})^2 = \pi \times 10^{-4}\,\text{cm}^2$

Thus

$$V_b = \frac{A_c K}{\gamma_{Hg}} = \frac{\pi(10^{-4}\,\text{cm}^2)\,0.2\,\text{cm/°C}}{1.82 \times 10^{-4} \frac{\text{cm}^3}{\text{cm}^3\,°C}} = 0.345\,\text{cm}^3$$

1.6 Find the spring scale (Fig. 1.11a) transfer function when the mass is negligible. Equation 1.24 becomes

$$B \frac{dy(t)}{dt} + K_s y(t) = x(t)$$

when $M = 0$. This is a first order system with

$$K = \text{static sensitivity} = \frac{1}{K_s}$$

$$\tau = \text{time constant} = \frac{B}{K_s}$$

Thus the operational transfer function is

$$\frac{y(D)}{x(D)} = \frac{1/K_s}{1 + \frac{B}{K_s}D} = \frac{1}{K_s + BD}$$

and the sinusoidal transfer function becomes

$$\frac{y(j\omega)}{x(j\omega)} = \frac{1/K_s}{1 + j\omega \frac{B}{K_s}D} = \frac{1/K_s}{\sqrt{1 + \frac{\omega^2 B^2}{K_s^2}}} \quad \angle \phi = \tan^{-1}\left(-\omega \frac{B}{K_s}\right)$$

1.7

$$a\frac{dy}{dt} + bx + c + dy = e\frac{dy}{dt} + fx + g$$

$$(a-e)\frac{dy}{dt} + dy = (b+f)x + (g-c)$$

This has the same form as equation 1.15 if $g = c$.

$$(\tau D + 1)y = Kx$$

$$\left[\frac{a-e}{d}D + 1\right]y = \frac{b+f}{d}x$$

Thus $\tau = \frac{a-e}{d}$

Basic Concepts of Medical Instrumentation

1.8 For a first order instrument

$$\frac{Y(j\omega)}{X(j\omega)} = \frac{K}{(j\omega\tau + 1)}$$

$$\left|\frac{K}{(j\omega\tau + 1)}\right| = \frac{K}{\sqrt{\omega^2\tau^2 + 1}} = 0.93 K$$

$$\left(\omega^2\tau^2 + 1\right)(0.93)^2 = 1$$

$$f = \frac{1}{2\pi}\omega = \frac{1}{2\pi}\sqrt{\frac{1 - (0.93)^2}{(0.93)^2 (.02)^2}} = 3.15 \text{ Hz}$$

$$\phi = \tan^{-1}(-\omega\tau) = -21.6°$$

1.9

$$y(t) = K \frac{Ke^{-\zeta\omega_n t}}{\sqrt{1 - \zeta^2}} \sin\left(\sqrt{1 - \zeta^2}\,\omega_n t + \phi\right) \text{ where } \phi = \sin^{-1}\sqrt{1 - \zeta^2}$$

$\zeta = 0.4;\ f_n = 85$ Hz

$$t_n = \frac{\frac{3\pi}{2} - \phi}{\omega_n \sqrt{1 - \zeta^2}} = 7.26 \text{ ms} \qquad t_{n+1} = \frac{\frac{7\pi}{2} - \phi}{\omega_n \sqrt{1 - \zeta^2}} = 20.1 \text{ ms}$$

$$y(t_n) = 10 + \frac{10}{\sqrt{1 - \zeta^2}} e^{-\zeta\omega_n t_n} \qquad y(t_{n+1}) = 10 + \frac{10}{\sqrt{1 - \zeta^2}} e^{-\zeta\omega_n t_{n+1}}$$

$$= 12.31 \qquad\qquad\qquad\qquad = 10.15$$

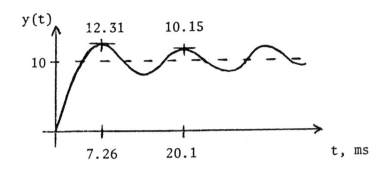

1.10. At the maxima y_n, y_{n+2}, y_{n+4}: $\sin(\ ) = -1$ at $\dfrac{3\pi}{2}, \dfrac{7\pi}{2}, \dfrac{11\pi}{2}$

and $\sqrt{1-\zeta^2}\,\omega_n t_n + \phi = \dfrac{3\pi}{2}$ $\qquad t_n = \dfrac{\dfrac{3\pi}{2} - \phi}{\omega_n \sqrt{1-\zeta^2}}$

at the minima $y_{n+1}, y_{n+3}\ldots$: $\sin(\ ) = +1$ at $\dfrac{5\pi}{2}, \dfrac{9\pi}{2}\ldots$

and $\sqrt{1-\zeta^2}\,\omega_n t_{n+1} + \phi = \dfrac{5\pi}{2}$ $\qquad t_{n+1} = \dfrac{\dfrac{5\pi}{2} - \phi}{\omega_n \sqrt{1-\zeta^2}}$

then form the ratio

$$\dfrac{y_n}{y_{n+1}} = \dfrac{\dfrac{K}{\sqrt{1-\zeta^2}}\,e^{-\zeta\omega_n t_n}}{\dfrac{K}{\sqrt{1-\zeta^2}}\,e^{-\zeta\omega_n t_{n+1}}} = \exp\dfrac{-\dfrac{3\pi}{2} - \dfrac{5\pi}{2}}{\sqrt{1-\zeta^2}} = \exp\dfrac{+\pi\zeta}{\sqrt{1-\zeta^2}}$$

$$\Gamma = \ln\dfrac{y_n}{y_{n+1}} = \dfrac{\pi\zeta}{\sqrt{1-\zeta^2}}$$

Solve for ζ

$$\zeta = \dfrac{\Gamma}{\sqrt{\pi^2 + \Gamma^2}}$$

Chapter 2
Basic Sensors and Principles

Robert A. Peura and John G. Webster

2.1 Let the wiper fraction $F = x_i/x_t$

$$v_o/v_i = \frac{R_m \| F R_p}{R_m \| F R_p + (1-F)R_P}$$

$$= \frac{\dfrac{R_m F R_p}{R_m + F R_p}}{\dfrac{R_m F R_p}{R_m + F R_p} + (1-F)R_P}$$

$$= \frac{1}{1 + \dfrac{R_m F R_p}{R_m + F R_p}(1-F)R_P}$$

$$= \frac{1}{\dfrac{R_m F + R_m + F R_p - F R_m - FFR_p}{R_m F}}$$

$$= \frac{1}{\dfrac{1 + F R_p/R_m - FFR_P/R_m}{F}}$$

$$= \frac{1}{\dfrac{1}{F} + \dfrac{R_p}{R_m}(1-F)}$$

Let $\varepsilon = R_p/R_m$

error $= F - v_o/v_i$

$$= F - \frac{1}{1/F + \varepsilon(1-F)}$$

$$= F - \frac{F}{1 + \varepsilon F - \varepsilon F^2}$$

$$= F - (F)(1 + \varepsilon F - \varepsilon F^2)^{-1}$$

$$d/dF(\text{error}) = 0 = 1 - (1)(1+\varepsilon F-\varepsilon F^2)^{-1} - (F)(-1)(1+\varepsilon F-\varepsilon F^2)^{-2}(\varepsilon - 2\varepsilon F)$$

multiply by $(1+\varepsilon F-\varepsilon F^2)^2$

$$0 = (1+\varepsilon F-\varepsilon F^2)^2 - (1+\varepsilon F-\varepsilon F^2) + (F)(\varepsilon - 2\varepsilon F)$$

expand, ignoring terms of $\varepsilon^2, \varepsilon^3, \ldots$

$$0 = 1 + 2\varepsilon F - 2\varepsilon F^2 - 1 - \varepsilon F + eF^2 + \varepsilon F - 2\varepsilon F^2$$

$$0 = -3\varepsilon F^2 + 2\varepsilon F = (\varepsilon)(F)(2-3F)$$

$$F = 0, 2/3$$

$$\text{error} = 0.67 - \frac{1}{1/0.67 + \varepsilon(1 - 0.67)}$$

$$= 0.67 - \frac{1}{1.5 + 0.33\varepsilon}$$

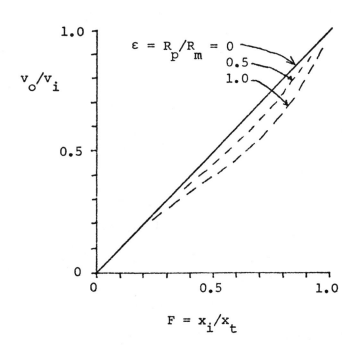

$$= \frac{0.22\varepsilon}{1.5 + 0.33\varepsilon} \approx 0.15\varepsilon = 0.15\, R_p/R_m$$

2.2 The resolution of the translational potentiometer is 0.05 to 0.025 mm. The angular resolution is a function of the diameter, D, of the wiper arm and would = (translational resolution/πD) × 360°. In this case the resolution is 2.87/D to 5.73/D degrees where D is in mm.

A multiturn potentiometer may be used to increase the resolution of a rotational potentiometer. The increased resolution is achieved by the gearing between the shaft whose motion is measured and the potentiometer shaft.

2.3. The elastic-resistance strain gage is nonlinear for large extensions (30%), has a dead band linearity due to slackness and is subject to long-term creep. Continuity in the mercury column and between the column and electrodes may be a problem. The gage has a high temperature drift coefficient. The dynamic response and finite mechanical resistance may cause distortion. These problems may be minimized by carefully selecting the proper size gage for the extremity. The gage should be slightly extended at minimum displacement when applied to eliminate the slackness problem. Mercury continuity checks may be made using an ohmmeter. The temperature drift problems may be minimized with continual calibration or by making measurements in a controlled temperature environment.

2.4 From (2.21)

$$E = 38.7T + (0.082/2)T^2 = 38.7T + 0.041T^2$$

T	38.7T	0.41T^2	E
°C	µV	µV	µV
0	0	0	0
10	387	4	391
20	774	16	790
30	1161	37	1196
40	1548	66	1614
50	1935	102	2037

The second term is small. The curve is almost linear but slightly concave upward.

2.5 From (2.22)

$$K = dE/dT = a + bt = 38.7 + 0.082T \; \mu V/°C$$

$$= 38.7 + 0.082(37) = 41.7 \; \mu V/°C$$

Basic Sensors and Principles   11

2.6   From (2.24)
$$\alpha = -\beta/T^2 = \frac{-4000}{(300)^2} = -4.4\%/K$$

2.7   There is always a voltage induced in each secondary, because it acts as the secondary of an air-core transformer. This voltage increases when the core is inside it.

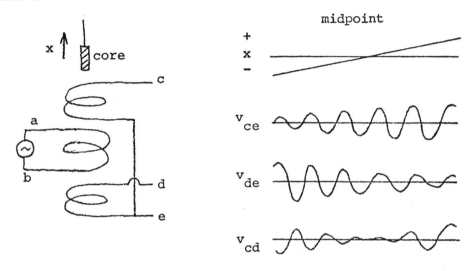

2.8   From (2.8) we can calculate x given the value of C.

$$C = \varepsilon_o \varepsilon_r A/x$$

C may be found from the corner frequency relationship

$$f = 1/2\pi RC = 20 \text{ Hz}$$

or

$$C = 80 \text{ pF}$$

Thus $x = \dfrac{(8.854 \times 10^{-12})(1 \times 10^{-4})}{80 \times 10^{-12}} = 1.11 \times 10^{-5} \text{m} = 11.1 \text{ }\mu\text{m}$

2.9   In Example 2.3 C = 500 pF for the piezoelectric transducer. The amplifier input impedance = 5 MΩ.

$$F = 0.05 \text{ Hz} = \frac{1}{2\pi RC_{equivalent}}$$

Thus

$$C_{\text{equivalent}} = 0.637 \times 10^{-6} = C_{\text{piezoelectric}} + C_{\text{shunt}}$$

$$C_{\text{shunt}} = 0.636 \; \mu F$$

The sensitivity will be decreased by a factor of 1300 due to increase in the equivalent capacitance.

2.10 Calculate the voltage from $V = Q/C = 1 \; \mu C/1 \; nF = 1$ kV. Because this is too high, add a shunt capacitor $C_s = 1 \; \mu F$ to achieve 1.0 V. Allow for a gain of 10. To achieve low corner frequency, add shunt $R_s = 1/(2\pi f_c C) = 1/(2\pi \cdot 0.05 \cdot 1 \; \mu F) = 3.2$ MΩ. To achieve gain of +10, select $R_f = 10$ kΩ, $R_i = 1.11$ kΩ. To achieve high corner frequency, $C_f = 1/(2\pi f_c R_f) = 1/(2\pi \cdot 100 \cdot 10 \; k\Omega) = 160$ nF.

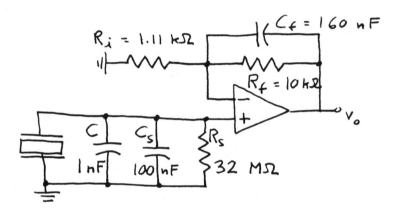

2.11 Select a feedback $C_f = 100$ nF (much larger than 500 pF). To achieve low corner frequency, add $R_f = 1/(2\pi f_c C_f) = 1/(2\pi \cdot 0.05 \cdot 100 \; nF) = 32$ MΩ. To achieve high corner frequency add separate passive filter or active filter with $R_o = 10$ kΩ and $C_o = 1/(2\pi f_c R_o) = 1/(2\pi \cdot 100 \cdot 10 \; k\Omega) = 160$ nF.

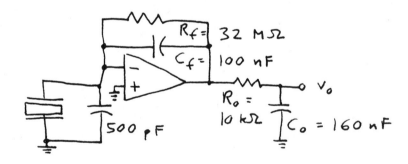

2.12 Typical thermistor V–i characteristics with and without a heat sink are shown below.

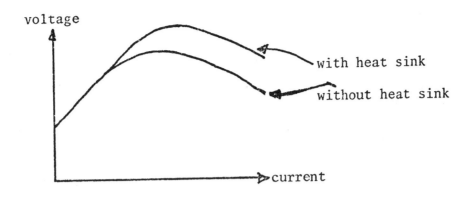

For low currents Ohm's law applies and the current is directly proportional to the applied voltage in both cases. The thermistor temperature is that of its surroundings. The system with a heat sink can reach higher current levels and still remain in a linear portion of the v–i curve since the heat sink keeps the thermistor at approximately the ambient temperature. Eventually the thermistor–heat sink combination will self heat and a negative-resistance relationship will result.

2.13 Assume $\varepsilon = 1.0$ and use (2.25).

$$W_\lambda = 37400/[\lambda^5(\exp(14400/\lambda 300) - 1)]$$
$$W_\lambda = 37400/[\lambda^5(\exp(48/\lambda) - 1)]$$
$$W_2 = 37400/[(32)(225 \times 10^6)] = 0.000005$$
$$W_5 = 37400/[3125)(15000)] = 0.0008$$
$$W_{10} = 37400/[100000)(120)] = 0.003$$
$$W_{20} = 37400/[3.2 \times 10^6)(11 - 1)] = 0.001$$

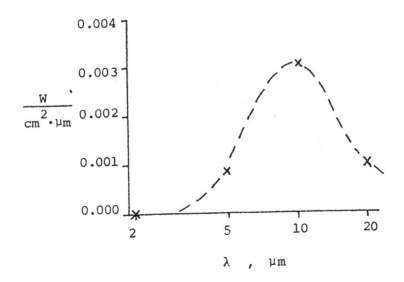

2.14 From (2.26),

$$\lambda_m = \frac{2898}{T} = \frac{2898}{300} = 9.66\ \mu m = 9660\ nm$$

From example 2.4

$$E_w = \frac{1240}{\lambda} = \frac{1240}{9660} = 0.128\ eV$$

2.15 Infrared and ultraviolet are passed better by mirrors because the absorption in the glass lenses is eliminated.

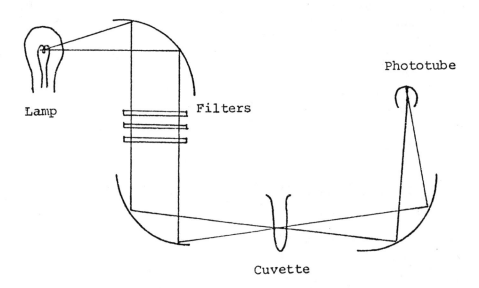

2.16 See section 2.16, photojunction devices. For small currents, beta, the current gain, increases with collector current. This produces the concave nonlinearity shown. Both nonlinearity and response time increase in the photo-Darlington because two transistors are involved.

Basic Sensors and Principles   15

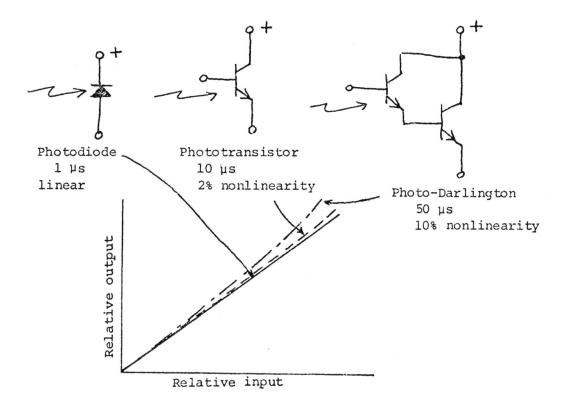

2.17 Try several load resistors as shown by the dashed load lines following. The maximum power is 2.5 µW. The load resistor, R = V/I = (0.5 V)/(5 µA) = 100 kΩ.

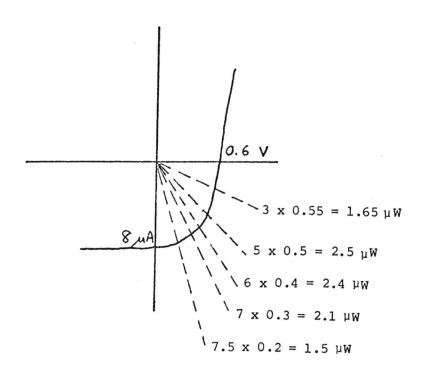

16 Basic Sensors and Principles

2.18 (a) shows the problem—the RC product is too high. (b) shows the simplest solution—the transistor input resistance is much lower than R. (c) shows that an op amp provides a virtual ground that provides a low input resistance. (d) shows that if R is divided by 10, the gain may be achieved by a noninverting amplifier. Active components must have adequate speed.

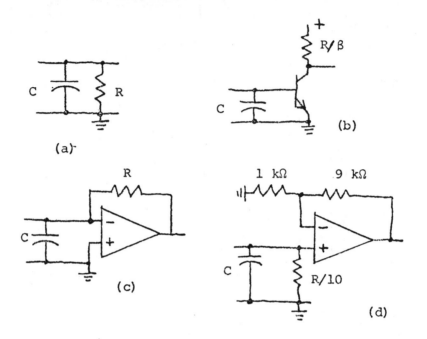

2.19 The solid line shows GaP. The dashed line shows Hb0. The dotted line shows CdS. The open circles show the relative combination product, which is calculated at each wavelength from the product of the other three curves.

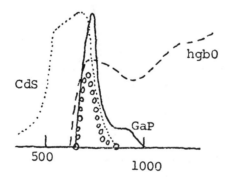

Chapter 3
Amplifiers and Signal Processing

John G. Webster

3.1(a) See Fig. 3.3(a)

$$v_i/v_o = -10 = -R_f/R_i = -R_f/20 \text{ k}\Omega$$

$$R_f = 200 \text{ k}\Omega$$

3.1(b) See section 3.13, Differential bias current

$$R = R_i \| R_f = \frac{R_i R_f}{R_i + R_f} = \frac{(20 \text{ k}\Omega)(200 \text{ k}\Omega)}{20 \text{ k}\Omega + 200 \text{ k}\Omega}$$

$= 18 \text{ k}\Omega$ placed between the positive input and ground.

3.1(c) See section 3.2, Summer
Choose $R_f = 100 \text{ k}\Omega$
Then design each input independently.

$$v_o/v_{i1} = -100 \text{ k}\Omega/R_{i1} = -10;\ R_{i1} = 10 \text{ k}\Omega$$

$$v_o/v_{i2} = -100 \text{ k}\Omega/R_{i2} = -2;\ R_{i2} = 50 \text{ k}\Omega$$

$$v_o/v_{i3} = -100 \text{ k}\Omega/R_{i3} = -0.5;\ R_{i3} = 200 \text{ k}\Omega$$

3.2 Gain = 20 V/150 mV = 133. To provide high input impedance, choose R_f = 13.3 MΩ, R_i = 100 kΩ. Then use Fig. E3.1 to make $v_o = 0$ when $v_i = -25$ mV (center of input range). $R_b = -R_i v_b/v_i = (100 \text{ k}\Omega)(15 \text{ V})/(-0.025 \text{ V}) = 60 \text{ M}\Omega$.

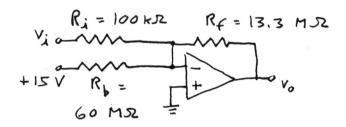

18 Amplifiers and Signal Processing

3.3 Gain = 20 V/200 μV = 100000. To provide high input impedance, choose R_f = 10 MΩ, R_i = 100 Ω. $R_b = -R_i v_b/v_i$ = (100 Ω)(15 V)/(−300 mV) = 5 kΩ. For smooth operation, the potentiometer should be smaller than R_b. The 100-Ω input impedance would require a buffer op amp to achieve high input impedance.

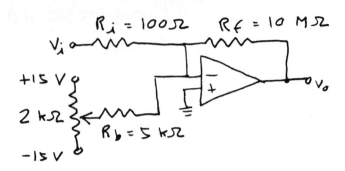

3.4 See Fig. 3.4(b)

$$v_o/v_i = (R_f + R_i)/R_i$$

$$10 = (R_f + 20\text{ k}\Omega)/20\text{ k}\Omega$$

$$R_f = 180 \text{ k}\Omega$$

See section 3.13, Differential bias current

$$R = R_i \| R_f = \frac{R_i R_f}{R_i + R_f} = \frac{(20\text{ k}\Omega)(180\text{ k}\Omega)}{20\text{ k}\Omega + 180\text{ k}\Omega}$$

$$= 18 \text{ k}\Omega \text{ placed between the positive input and } v_i$$

3.5 Modify Fig. 3.5(a) so the top two resistors have primes: R_3', R_4'. Connect v_3 to v_4 and drive with v_{cm}. Use superposition. The output due to v_4:

$$v_{o4} = v_{cm} \left[\left(\frac{R_4}{R_3 + R_4} \right) \left(\frac{R_3' + R_4'}{R_3'} \right) \right]$$

$$v_{o5} = v_{cm} \left[-\frac{R_4'}{R_3'} \right]$$

$$v_o = v_{o4} + v_{o5}$$

$$= v_{cm} \left[\left(\frac{R_4}{R_3+R_4}\right) \left(\frac{R_3'+R_4'}{R_3'}\right) - \left(\frac{R_4'}{R_3'}\right) \right]$$

Worst case: $R_3, R_4' = 105; R_3', R_4 = 95$

$$v_o/v_{cm} = \left[\left(\frac{95}{105+95}\right) \left(\frac{95+105}{95}\right) - \left(\frac{105}{95}\right) \right]$$

$$= 1 - 1.1 = -0.1$$

$$\text{CMRR} = G_d/G_c = -1/-0.1 = 10.$$

3.6 See Fig. 3.5(a). op amp loads should be > 2 kΩ, so choose 10 kΩ.

$$G_d = 5 = \frac{2R_2 + R_1}{R_1} = \frac{2R_2 + 10\text{ k}\Omega}{10\text{ k}\Omega}$$

$R_2 = 20$ kΩ

$v_o = (v_4 - v_3) R_4/R_3$

$$\frac{v_o}{v_4-v_3} = \frac{R_4}{R_3} = \frac{R_4}{10\text{ k}\Omega} = 6$$

$R_4 = 60$ kΩ

3.7 From (2.6), $\Delta v_o = v_i \Delta R/R = 5$ V(0.003) = 0.015 V. Gain = 20/0.015 = 1333. Assume R = 120 Ω. Then the Thevenin source impedance = 60 Ω. Use this to replace R_3 of Fig. 3.5(a) right side. Then $R_4 = R_3$(gain) = 60 Ω(1333) = 80 kΩ.

3.8 See Fig. 3.6. Voltage across R_3 is one-fourth of hysteresis width (2 V) or 0.5 V. Assume saturation at ±12.5 V. Choose $R_3 = 0.5$ kΩ, then $R_2 = 12$ kΩ.

Center of hysteresis window ($-v_{ref}$) is at +1 V.
Therefore $v_{ref} = -1$ V. Choose $R_1 = 10$ kΩ.

3.9 See Fig. 3.7(a) for the origin of the circuit. The difference between the op amp output and circuit output is caused by the diode voltage drop.

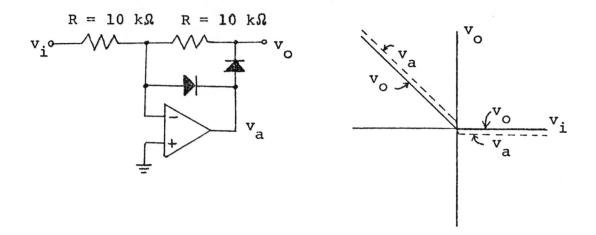

3.10 See Fig. 3.8. Maximum $V_{BE} = 0.66$ V. To obtain $v = 4$ V, the feedback ratio = $4/0.66 = 6$. Choose resistors > 2 kΩ. To obtain both polarities use two transistors as shown.

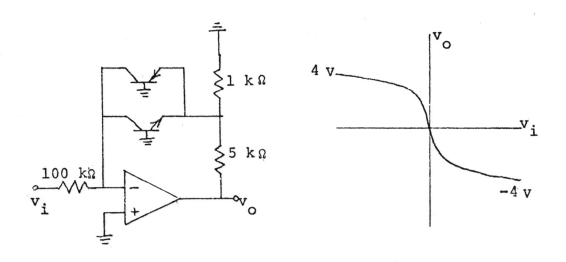

3.11   See Fig. 3.9

$$v_o = -\frac{1}{RC}\int v_i \, dt$$

$$\frac{dv_o}{dt} = -\frac{1}{RC} v_i$$

$$0.1 = -\frac{1}{RC} 10$$

$$RC = 0.1 \text{ s}$$

$$(1 \text{ M}\Omega)(0.1 \text{ μF}) = 0.1 \text{ s}.$$

3.12 $i = \dfrac{v}{R} = \dfrac{0.005}{1 \text{ M}\Omega} = 5 \text{ nA}$

$q = it = Cv$
$(5 \times 10^{-9} \text{ A})(t) = (10^{-7} \text{ F})(12 \text{ V})$
$t = \dfrac{12}{5} \times 10^2 = 240 \text{ s}$

Add offset voltage nulling pot to op amp, or add feedback resistance.

3.13 $i_R = 0$
$q = it = Cv$
$(200 \text{ pA})(t) = (0.1) \text{ μF} (12 \text{ V})$
$(0.002 \times 10^{-7} \text{ A})(t) = (10^{-7} \text{ F})(12 \text{ V})$
$t = \dfrac{12}{0.002} = 6000 \text{ s } (100 \text{ min}).$

Add 1 MΩ resistor between positive input and ground.

3.14 $v_o = -RC \, dv_i/dt$
$-10 \text{ V} = -RC \, (100 \text{ V/s})$
$RC = 0.1 \text{ s}$

See Fig. 3.12

Choose $R = 1 \text{ M}\Omega, C = 0.1 \text{ μF}$

3.15 See Fig. 3.12(b). The largest practical $C_i = 1$ μF.

$$\tau = R_i C_i = \frac{1}{2\pi f} = \frac{1}{2\pi(0.05)} = 3.2 \text{ s}$$

$$R_i = \frac{3.2}{10^{-6}} = 3.2 \text{ M}\Omega$$

$$R_f = 20(3.2) = 64 \text{ M}\Omega$$

This is very high and will cause large offsets because of bias currents. Therefore obtain gain from noninverting amplifier:

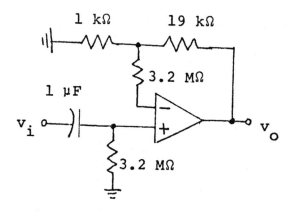

This is a first-order high-pass filter (section 1.9). $\frac{V_o}{V_i}$ 20 exp $(-t/\tau)$

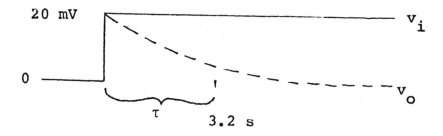

3.16 Start with a high-pass RC passive filter with input impedance $R = 10$ MΩ. $C = 1/(2\pi f_c R) = 1/(2\pi \cdot 10 \cdot 10 \text{ M}\Omega) = 1.6$ nF. Follow with Fig. 3.4(b) noninverting amplifier with $R_f = 10$ kΩ and $R_i = 1.1$ kΩ

3.17   See Fig. 3.12(c)

$$\frac{v_o}{v_i} = -\frac{Z_f}{Z_i} = \frac{1}{Z_i G_f} = -\frac{1}{\left(\dfrac{1}{j\omega C_i} + R_i\right)\left(j\omega C_f + \dfrac{1}{R_f}\right)}$$

$$= -\frac{(j\omega C_i)(R_f)}{(1 + j\omega R_i C_i)(j\omega R_f C_f + 1)}$$

$$= -\underbrace{\frac{R_f}{R_i}}_{\substack{\text{dc}\\\text{gain}}} \underbrace{\left(\frac{j\omega R_i C_i}{1 + j\omega R_i C_i}\right)}_{\substack{\text{hi-pass filter}\\\text{due to input}\\\tau_i = R_i C_i}} \underbrace{\frac{1}{1 + j\omega R_f C_f}}_{\substack{\text{low-pass filter}\\\text{due to output}\\\tau_i = R_i C_i}}$$

3.18   Use Fig. 3.12(c). $R_i = 10$ kΩ. $C_i = 1/(2\pi f_c R_i) = 1/(2\pi \cdot 1$ kHz$\cdot 10$ k$\Omega) = 16$ nF. For gain = 1, $R_f = 10$ kΩ. $C_f = 1/(2\pi f_c R_f) = 1/(2\pi \cdot 10$ kHz$\cdot 10$ k$\Omega) = 1.6$ nF.

3.19   Section 6.6 states 90 mV from 25 to 3000 Hz. Gain = 10 V/90 mV = 111. Use Fig. 3.12(c). Assume $R_i = 10$ kΩ. $C_i = 1/(2\pi f_c R_i) = 1/(2\pi \cdot 25$ Hz$\cdot 10$ k$\Omega) = 640$ nF. For gain = 111, $R_f = 11.1$ MΩ. $C_f = 1/(2\pi f_c R_f) = 1/(2\pi \cdot 3$ kHz$\cdot 11.1$ M$\Omega) = 5$ pF.

3.20   Gain-bandwidth product is 4 MHz. Therefore at 100 kHz, gain is limited to 40. Cascade two op amps with a gain of 10 to achieve a gain of 100 at 100 kHz. Each could have $R_i = 5$ kΩ, $R_f = 50$ kΩ.

3.21   Gain-bandwidth product is 4 MHz. At 100 Hz, op amp gain is 40000.

   (op amp gain) = (circuit gain) (loop gain)
   40000 = 1000 (loop gain)
   loop gain = 40

3.22   See Fig. 3.11. Break the feedback between v_o and R. The signal passes through R and C (to ground). This low-pass filter causes $-90°$ phase shift.

Then the signal passes through the op amp, which produces $-90°$ of additional phase shift. The total phase shift of $-180°$ causes oscillation as described in section 3.11.

24 Amplifiers and Signal Processing

3.23 Assume $R_i = 1 \text{ k}\Omega$, $R_f = 1 \text{ M}\Omega$

Inverter: $R_{ai} = R_i = 1 \text{ k}\Omega$

$R_{ao} = R_o/(\text{loop gain}) = 100/40 = 2.5 \text{ }\Omega$

Noninverting amplifier:

$R_{ai} = R_d (\text{loop gain}) = (1 \text{ T}\Omega)(40) = 40 \text{ T}\Omega$

$R_{ao} = R_o/(\text{loop gain}) = 100/40 = 2.5 \text{ }\Omega$

3.24 $q = it = Cv$

$(0.020 \text{ A})(10^{-6} \text{ s}) = C(15 \text{ V})$

$C = \dfrac{20}{15} \times 10^{-9} = 1.3 \times 10^{-9} \text{ F} = 1.3 \text{ nF}$

3.25 (a) See Fig. 1.6(a)

$2\pi fRC = 1$

$R = \dfrac{1}{2\pi 20 (10^{-7})} = 80 \text{ k}\Omega$

(b) See Fig. 3.12(a)

$C_F = 0.1 \text{ µF}$, $R_i = R_f = 80 \text{ k}\Omega$

3.26 See Fig. 3.17. With forward current flowing through D1 and D2, a higher voltage drop in D1 makes B more negative than A. With forward current flowing through D3 and D4, the voltage at C and A are equal. These two voltages appear at D and v_o. The low-pass filter averages these to produce a negative voltage one-half that seen at B.

3.27 Figure 1.6(a) shows a low-pass filter. The corner frequency $f_c = 1/(2\pi\tau) = 100$ Hz. Select a reasonable value capacitor $= 100$ nF. $\tau = RC$. Then $R = 1/(2\pi \cdot 100 \text{ Hz} \cdot 100 \text{ nF}) = 15.9 \text{ k}\Omega$.

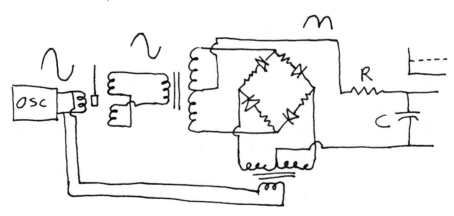

Chapter 4
The Origin of Biopotentials

John W. Clark

4.1 The four main factors involved in the movement of ions across the cell membrane in the resting-state are: (a) the ionic concentration gradients across the membrane, (b) the opposing electric field across the membrane (the membrane, acting as a charge separator), (c) permeability for a specific ionic species, and (d) active $Na^+ - K^+$ transport which is responsible for the maintenance of the ionic imbalance between the internal and external media of the cell in the steady-state.

4.2 Use (4.1), $E_K = 0.0615 \log_{10}[K]_o/[K]_i = 0.100$.

$0.100/0.0615 = 1.626 = \log_{10}[K]_o/[K]_i$

$10^{1.626} = 42.27 = [K]_o/[K]_i$

$[K]_i = [K]_o/42.27 = 4/42.27 = 0.095$ mmol/l

$[Na]_i = [Na]_o/42.27 = 145/42.27 = 3.43$ mmol/l

$[Cl]_i = 42.27[Cl]_o = 42.27 \cdot 120 = 5072$ mmol/l (negative ion)

4.3 This problem is best answered with the aid of the Fig. P4.3. Here the current application circuit consists of an intracellular micropipet and an extracellular electrode placed on the outer membrane surface and these electrodes are connected to the terminals of a stimulus isolation unit (SIU). We will assume that the polarity of the terminals of the SIU can be changed at will via a switch available on the unit. We will also assume that a recording electrode system consisting of a second micropipet and extracellular reference electrode can be placed some distance from the stimulating site for the purposes of detecting gross excitability changes in the cell (i.e. to determine whether an action potential occurs or not).

With the passage of current through the membrane via stimulating electrode pair, an ohmic potential drop is produced across the membrane that either increases or decreases the magnitude of the steady resting potential, by the amount of the potential drop. Whether membrane potential is increased or decreased simply depends on the polarity of applied voltage. With current flowing outward through the membrane (i.e., with the micropipet as the anode and the extracellular electrode as the cathode) the potential drop has a polarity opposite to that of the resting transmembrane potential and therefore the magnitude of the resting potential decreases (this is called depolarization). With current flowing into the membrane, the ohmic potential drop is of similar polarity to that of the resting transmembrane potential and the magnitude of the resting potential is increased (hyperpolarization).

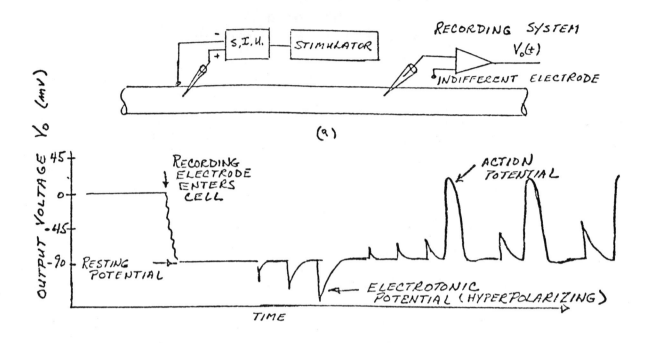

Fig. P4.3

Figure P4.3 illustrates the following:

(1) Prior to penetration of the membrane the potential difference between the micropipet and extracellular electrode is zero. After successful penetration, the resting potential is established.

(2) Brief hyperpolarizing stimuli (micropipet negative) simply produce nonpropagating electronic potentials that increase the magnitude of the resting potential without eliciting an action potential.

(3) Brief depolarizing stimuli (micropipet positive) produce nonpropagating electronic potentials that decrease the magnitude of the resting potential. With sufficient membrane depolarization a threshold level in membrane potential is exceeded and an action potential is elicited. Stronger depolarizing stimuli have no effect on the action potential waveshape demonstrating the "all or none" behavior of the action potential.

4.4 A diagram of the lines of current flow between the external stimulating pair of electrodes is shown in Fig. P4.4(a). Current flows outward through the membrane under the cathode and inward through the membrane under the anode. Thus, as explained in the previous example, membrane potential is depolarized under the cathode and hyperpolarized under the anode. The anodal and cathodal electrotonic potentials are also illustrated in Fig. P4.4(b). These electronic potentials are recorded by positioning the exploring electrode, for example, in the immediate vicinity of the anode, and then the cathode.

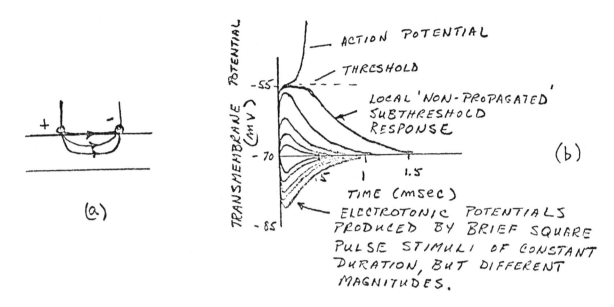

Fig. P4.4

4.5 Figure 4.3 shows that each section of the distributed parameter model forms an R–C low-pass filter. Multiple sections form multiple low-pass filters. Thus the response due to the stimulating square-wave pulse is progressively smoothed and attenuated as the separation distance increases.

4.6 The nerve fiber action potential is a brief transient depolarizing disturbance in the transmembrane potential (V_m) of the cell that is conducted in an unattenuated fashion along the length of the fiber. Assuming that an action potential is generated, the membrane charge distribution is roughly that shown in Fig. 4.4 of the text. The direction of propagation is to the left in this figure and lines of "local circuit" current flow precede the discrete "active source region" depolarizing the inactive resting membrane segments lying ahead of the active region. Those lines of current flow emanating in the active source region and traveling backward into the regions recently occupied by the active region, do not re-excite the membrane because it is in an absolute- or relatively-refractory state. The depolarizing current preceding the source region and gradually bringing successive increments of membrane into the active state by exceeding the membrane potential threshold, is primarily carried by Na^+. The repolarization current flowing from the source region into the previously activated region is primarily carried by K^+.

4.7 Stimulating a fiber with a closely-spaced external stimulating pair of electrodes located in the middle of the fiber would cause action potentials, traveling in opposite direction, to be conducted to both ends of the fiber. The action potential elicited at the cathode of the stimulating pair would occur first in time since the membrane region lying underneath the cathode is depolarized upon application of a brief stimulus. The membrane region beneath the anode is hyperpolarized initially, but this inhibitory effect is eventually overcome by depolarizing local circuit current flow from the active membrane source region in the vicinity of the cathode. The hyperpolarizing background provided by the anode produces a form of temporary block to neural transmission in the vicinity of the anode known as "anodal" block. However, the strong depolarizing currents from the source region overcome this anodal block and an action potential is eventually launched from the anodal region. This is represented schematically in the figure below. Of course, there will be a small temporal difference in the time of occurrence of the two action potentials due to the time required to overcome anodal block. One can easily imagine other configurations of stimulating electrodes, and therefore the reader is encouraged to answer this question using the same principles applied to other possible configurations.

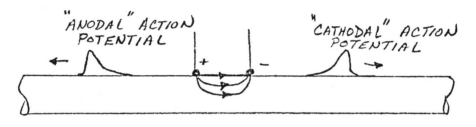

Fig. P4.7

4.8   See text.

4.9   (a)   The field potential is attenuated and loses frequency content as radial distance from the nerve trunk increases. This is discussed in section 4.2 and further information on this topic is provided in a paper by Greco, E.C. and Clark, IEEE Transactions on Biomedical Engineering 25, 18–23, 1977.

(b)   The bioelectric source has been shown experimentally to behave approximately as a constant action current source. This means that it can deliver the same current associated with the action potential, to a wide variety of resistive loads presented by the extracellular volume conductor. Thus, increasing the specific resistivity ρ of the bathing medium increases the total external resistance seen by the "constant current" bioelectric source. consequently, via Ohm's law applied to the volume conductor, the field potential magnitude increases with increasing values of ρ. An increase in ρ can be achieved experimentally by mixing various amounts of Ringer fluid and an isotonic dextrose (sugar) solution. For example, the specific resistivity of a solution containing two parts isotonic dextrose solution to one part amphibian Ringer solution, is more than four times the specific resistivity of the Ringer solution alone.

(c)   Changing the radius changes the total bathing medium resistance and thereby affects the field potential in predictable manner. That is, a decrease in radius, increases the external resistance which increases the magnitude of the field potential.

(d)   A volume conductor of finite dimensions can be considered an essentially "infinite" volume conductor when the dimensions of the conductor are large relative to the field extent of the bioelectric source. The adequacy of this assumption is proven experimentally by probing the extracellular field by means of an appropriate exploring electrode.

4.10   In this problem we might imagine the median nerve in the arm as the bioelectric source imbedded in a volume conductor that is roughly cylindrical in cross-section but gently tapered from the forearm to the wrist. As a first approximation we could consider the resistive volume conductor medium to be uniform, isotropic and homogeneous, and characterized by an average bulk specific resistivity ρ_e. The arm is comprised of many different tissues of different specific resistivities (e.g., extracellular fluid, blood, connective tissue, skeletal muscle, tendon, bone. We assume an average figure for the specific resistivity of the extracellular medium and label it $\bar{\rho}_e$. Considering the spatial extent of the active source region small relative to axial dimension of the volume conductor, we can easily interpret the experimental recordings shown in Fig. 4.8 in terms of a change in the radial dimensions of the volume conductor. In a qualitative sense, the nerve acts as a constant action current source and the extracellular medium at the wrist provides a more resistive loading medium for the current source, than the extracellular medium at the level of the forearm. Hence the observed surface potential is larger at the wrist and its frequency content is greater (i.e., the medium acts as a low-pass filter and the greater the radial extent of this medium, the greater the degree of low-pass filtering). We could also assume that there are regional differences in the average specific resistivity $\bar{\rho}_e$, since the extracellular medium at the wrist contains larger connective tissue, tendon and bone components than the extracellular forearm medium. These tissues have a higher specific resistivity than blood, skeletal muscle, etc., and hence $\bar{\rho}_e$ at the wrist is conceivably much greater than that at the forearm. This difference is in a direction to further enhance the magnitude of the recorded surface potential at the wrist.

4.11   See text section 4.4, Reflexly Evoked Potentials, Fig. 4.9.

4.12   Localized differences in the conduction velocities of component nerve fibers of a motor nerve innervating a particular skeletal muscle, results in a desynchronization of the neural volley of impulses that ultimately causes that muscle to contract. The electroneurogram and the electromyogram will exhibit this desynchronization (e.g., Fig. 4.12(b)) and if this desynchronization is pronounced, the mechanical contraction of the muscle will be uncoordinated and spastic.

4.13   Literature search question. Explained in text.

4.14   Literature search question.

4.15 (a) Internodal tracts: three bundles (anterior, middle and posterior) of specialized conducting tissue running between the SA and AV nodes. See Fig. 4.13 and 4.14.

(b) Subendocardial layer: the layer of cardiac musculature lying immediately beneath the endocardium (the endothelial lining of the ventricular penetrate the main cardiac musculature (the myocardium) of the ventricular wall. The extent of this penetration may be as large as 30% of the wall thickness in some species although only 15–20% in man.

(c) Intercalated disk: A region of close abutment between adjacent cardiac cells (see Fig. 4.15).

(d) Bundle branches: The specialized conduction system bifurcates into two (right and left) bundle branches in the uppermost portions of the intraventricular septum near the atrioventricular border of the heart. See Fig. 4.13 and 4.14.

(e) Ventricular activation: The sequence of events associated with the electrical activation of the ventricle. This sequence is illustrated in Fig. 4.16.

4.16 All three of the Einthoven leads (I, II and III) have a similar wave shape and the R wave is a positive deflection when the recording polarity conventions are observed. The lead II waveform is illustrated below where the P wave indicates the onset of atrial activity, the QRS complex indicates the onset of ventricular activation and the T wave results from ventricular repolarization. Atrial repolarization occurs more slowly than ventricular repolarization and the field potential associated with it is small and masked by the QRS complex. The P–R interval is measured from the onset of the P wave to the onset of the QRS complex (in some cases the ECG will show no Q wave, hence the name "P–R interval"). The majority of the P–R interval is contributed by conduction delay in the AV node; this has the effect of separating the mechanical contraction of the atria and ventricles. Of course, atrial conduction time and the time taken for conduction along the bundle branches to the ventricles are also contributing factors to the P–R interval. The S–T segment duration is directly related to the duration of the plateau region of the action potential recorded via micropipet from the typical ventricular cell (see Fig. 4.14). Since this portion of the normal ventricular action potential is constant one would expect the S–T segment to be zero. In various forms of cardiac disease the time course of potential in the plateau is not held constant and the resulting S–T segment is no longer isoelectric.

4.17 Literature search question. A typical His bundle recording obtained via intracardiac catherization is shown below where A is the atrial electrogram, H the His bundle response and V the ventricular electrogram. The technique is very useful in the study of conduction system disorders.

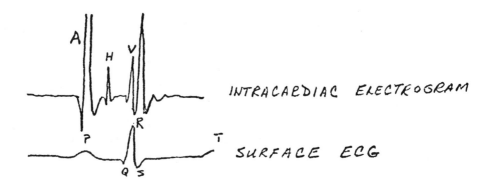

4.18 The ventricular action potential has a long refractory period so that the mechanical response of the ventricle will be discrete; one coordinated mechanical response per electrical activation sequence. Otherwise, with additional electrical stimuli, the mechanical responses would summate producing a prolonged contraction.

4.19 A PVC occurs early. The algorithm should test that the $R-R_{t-1}$ interval < 0.8 AR–R_{t-2}. (AR–R = average of 10). A PVC is followed by a compensatory pause. The algorithm should test that $R-R_{t-1} + R-R_t \approx 2(AR-R_{t-2})$. A PVC is wide. The algorithm should test that width $W > 1.3AW$. All three tests should be positive.

4.20 A rudimentary block diagram of the retina considered as a photoelectric sensor is shown below.

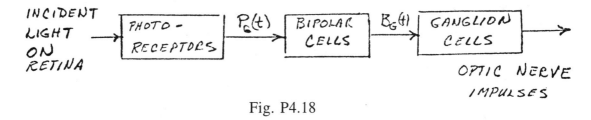

Fig. P4.18

The output of the photoreceptor unit P_G is a slow wave response similar to that of a generator potential. The output B_G of the bipolar cell layer is a slow wave as well. However, the output of the ganglion cell layer is a neural impulse train. Retinal electrophysiology is much more complicated than the brief description given in the text and the interested reader should be directed to Chapter 15 of Mountcastle (1980). Note that only the main transmission pathway is indicated in this diagram and that the roles played by the horizontal and amacrine cells of the retina are not indicated.

4.21 In terms of retinal cellular activity, the a-wave of the ERG is contributed by cells of the photoreceptive layer, the b-wave is contributed by bipolar and ganglion cells while the c-wave is contributed by cells of the pigment epithelium. for a detailed treatment of this subject see Mountcastle (1980), Chapter 15.

4.22 In terms of volume conductor theory the eye may be considered to be a spherical volume conductor with a sheet-like bioelectric source (the retina) covering most of one hemisphere; the corneal surface is considered to lie opposite this surface and potential measurements are conveniently made on this surface via placement of a contact lens with embedded measurement electrode. A brief flash of light presented to the dark-adapted eye will evoke a typical flash ERG which is recorded as the potential difference between the contact lens electrode and an indifferent electrode as in Fig. 4.25. The discussion included in the chapter section entitled "Spatial Properties of the ERG" indicates that it is also possible with the utilization of averaging techniques to record ERGs from localized regions of the retina. These regions have each been subjected to a brief low-intensity localized light stimulus against a general steady background illumination that provides partial adaptation of the entire retina. In the dark-adapted eye, there is a great deal of lateral communication between retinal units and a localized light stimulus will provide a more diffuse response from cells lying outside the boundary of the area stimulated. This lateral intercommunication between retinal units is inhibited with relatively high-intensity background illumination of the retina. Relatively low-intensity localized light stimuli are usually preferred which produce relatively low-amplitude ERG signals (the background illumination partially bleaches the photoreceptors, particularly the rods, thus diminishing the magnitude of the recorded ERG). These low-amplitude signals are detected in the presence of measurement noise via signal averaging. In this manner ERG responses from several regions of the retina can be obtained via measurement at a fixed corneal recording site. The text also cites literature demonstrating the linear superimposition of these local ERG responses. The reason for employing only a single recording site in this study is that it has been experimentally demonstrated by Krakau (1958) that the field potential within the volume conductor drops off rapidly in the vicinity of the thin retinal source, and then very slowly at larger distances from the source such as in the vicinity of the corneal surface. Thus, recording potentials at a variety of corneal and nearby scleral measurement sites will produce a data set that is very similar indeed. It follows that this data set would be of little value in solving an inverse field problem for the purpose of identifying the retinal bioelectric sources.

A gross model of the <u>retinal source</u> (mainly for purposes of characterizing b-wave activity in the ERG) would consist of a mosaic of nonuniform retinal regions each characterized electrically by a dipole current source per unit volume.

The local ERG response obtained via application of a localized light stimulus against a general background illumination, allows the experimenter to excite the dipoles of the model individually. Processing and analysis of the resultant ERG data, could result in the specification of a set of retinal current dipole parameters for an individual subject. This equivalent-source characterization of the retinal biopotential generator could have diagnostic significance.

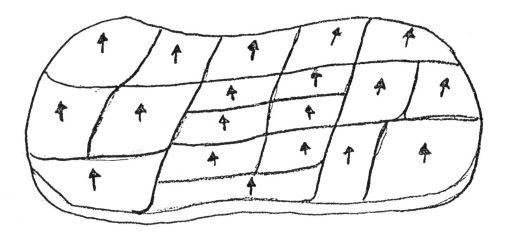

Fig. P4.22   Distribution of Retinal Dipole Sources

4.23   The subject of the use of the corneal-retinal potential in the measurement of eye movements is discussed in the section entitled "The Electro-oculogram (EOG). Briefly, there is an approximately linear relationship between the horizontal angle of gaze and EOG output up to approximately ±30 degrees of arc. The measurements do however suffer from a lack of accuracy at the extremes of this range; that is, eye movements of less than one or two degrees are difficult to record in the presence of noise compounded of effects from ongoing EEG and EMG activity as well as amplifier noise. Large eye movements (> 30 degrees of arc) do not produce bioelectric signal amplitudes that are strictly proportional to eye position. For greater detail regarding the accuracy of this technique, consult North (1965) and Kris (1960). The EOG is frequently the method of choice for recording eye movements in sleep and dream research, in recording eye movements from infants and children, and in evaluations of reading ability and visual fatigue.

4.24   The functional role played by each of these CNS structures is given in the text in section 4.8. Additional brief descriptions are given below.

(a)   The ascending general sensory nerve fiber pathways to the brain are input pathways conveying general sensory information regarding pain, temperature, coarse touch and pressure, fine touch, proprioceptive and kinesthetic information regarding the location of the body in space, etc. These general sensory pathways are "hard-wired" to an amazing degree and are divided not only according to the part of the body from which they originate but also according to the specific sensory modality (pain, temperature, etc.) that they mediate.

(b)   The RAS is discussed specifically under the subheading <u>Reticular Formation</u> in section 4.8. Activity of the RAS may be considered essential for the existence of a state of "awareness." It is a complex polysynaptic network within the reticulate formation that sends projecting fibers to the thalamus (the specific thalamocortical fibers to the cortex) and other fiber bundles that bypass the thalamus and project diffusely to the cortex (nonspecific thalamocortical fibers). The RAS is capable of influencing the electrical activity of the cortex to a remarkable degree.

(c) The precentral gyrus is primary motor cortex of the brain, and large fast-conducting nerve fibers originate in this region from large pyramidal cells called Betz cells; the fibers descend to the spinal cord in a well-defined tract called the corticospinal tract or pyramidal tract. Their objective is to make contact with neuronal elements of the ventral horn of the spinal cord at various spinal levels either by direct synaptic contact, or more frequently indirectly via interneurons.

The postcentral gyrus is the primary cortical projection area for the general sense pathways mentioned in part (a) of this question. It, like the incoming general sense pathways, is laminated typographically according to body position and sensory modality.

(d) The primary auditory cortex is the primary cortical projection area for the auditory system and is located on the temporal lobe as can be seen in Fig. 4.28. Similarly the primary visual cortex is the primary projection area for the visual system and is located on the occipital cortex of the brain (see Fig. 4.28).

(e) Activation of the specific thalamocortical pathway to the brain elicits a specific, localized cortical response that is in fact specific for a given type of sensory modality. This response to general sensory stimulation is often called a "primary" cortical response. Activation of the nonspecific thalamocortical pathways involves activation of the RAS (which by its nature is nonspecific) and the evoked cortical response is quite diffuse. This latter response is frequently called the diffuse "secondary" cortical response to general sensory stimulation and it can be recorded from several areas of the cortex simultaneously.

With general sensory stimulation, as in the case of stimulation of the sciatic nerve of the leg, a primary evoked cortical response is recorded in the vicinity of the postcentral gyrus, followed by a slower, larger-amplitude secondary response. The primary response is recorded only in a localized region of the postcentral gyrus, while the secondary response may be recorded over wide areas of the cortex.

4.25 This subject matter is discussed in the subsection entitled <u>Electrical Potentials from the Brain</u> in section 4.8. Briefly, slow-wave activity from the brain largely originates from cortical surface layers, particularly layer I which is rich in input fibers, dendrites from pyramidal cells, etc. There is a large nonspecific thalamocortical input to this level, whose influence is demonstrated particularly during synchronized slow-wave activity of the brain. The large pyramidal cells also constitute current sources embedded within the cortical substance and could possibly contribute to the general field potential as well. However, the potential in the vicinity of an individual soma is greatest and the field falls off fairly rapidly with increasing distance from the soma. This source of bioelectric activity consisting of the combined field potential of the giant pyramidal cells of layer V of the cortex, lies more distant from a scalp electrode recording site, compared with the dense neural mat comprising layer I of the cortex. Hence it has a smaller influence on recorded potential. Some of the smaller pyramidal cells of layers II and III could also possibly contribute to the general field potential recorded by a scalp macroelectrode or a cortical surface electrode in the case of an electrocorticoencephalogram. See Figs. 4.29 and 30.

4.26 In general terms, the problem involves the construction of a set of bandpass filters with appropriate center frequencies and bandwidths set for each of the major slow-wave components of the EEG. The EEG is presented simultaneously to the parallel set of bandpass filters to accomplish the spectral detection of alpha, beta, theta and delta wave activity.

   alpha: 8–13 Hz
   beta: 14–30 Gz
   theta: 4–7 Hz
   delta: <3.5 Hz

A refinement of this approach would allow us to distinguish between beta I and beta II EEG activity, as well as slow wave components in the EEG waveform other than the four standard waves discussed above.

4.27 This subject is discussed adequately in the text in the subsection entitled "The Volume Conductor Problem in Electroencephalography" in section 4.8. It is also a literature search question.

4.28 Every differential amplifier has 3 inputs. One of these must be connected to an electrode and to ground to keep the differential inputs within the common mode voltage range and to lead interference currents from the body to ground. Other connections shown.

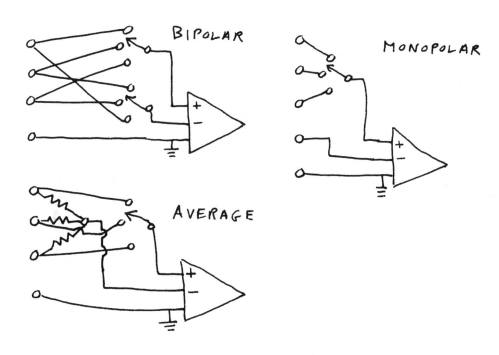

4.29 Three electrodes over the occipital lobe detect the 100-µV EEG and feed a differential amplifier with a gain of 10000. A band-pass filter centered at 10 Hz selects the alpha waves, which are demodulated and filtered to yield a dc value proportional to amplitude. This modulates the frequency of an acoustic tone, and the subject attempts to maximize the frequency.

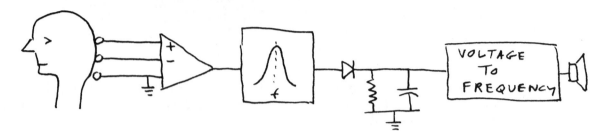

Chapter 5
Biopotential Electrodes

Michael R. Neuman

5.1 The reaction

$$Ag \rightarrow Ag^+ + e^- \qquad (P\ 5.1.1)$$

gives one silver ion per electron. By definition, a current of 10 µA is equivalent to

$$10\ \mu A = 10^{-5}\ C/s \qquad (P\ 5.1.2)$$

The charge on the electron is 1.6×10^{-19} C. Thus the number of silver ions oxidized at the electrode-electrolyte interface per second will be

$$N = \frac{10^{-5}}{1.6 \times 10^{-19}} = 6.25 \times 10^{13} s^{-1} \qquad (P\ 5.1.3)$$

5.2 Use Fig. 5.2.
Anode reaction: $Ag + Cl^- \Rightarrow AgCl\downarrow + e^-$
Cathode reaction: $2H^+ + 2e^- \Rightarrow H_2 \uparrow$.

5.3 Use Eq. (E5.6): $N = 0.15\ C/(1.6 \cdot 10^{-19}\ C/atom) = 9.4 \cdot 10^{17}$ atoms.
Use Eq. (E5.7): $N = 9.4 \cdot 10^{17}/(6.03 \cdot 10^{23}) = 1.56 \cdot 10^{-6}$ mol.
Weight of AgCl = $142.3 \cdot 1.56 \cdot 10^{-6} = 222 \cdot 10^{-6}$ g.
Suggest electrode of Fig. 5.16(c) with 1 cm^2 area.

5.4 The equivalent circuit model of an electrode with electrolyte gel coupling it to the skin is that shown in Fig. 5.4. When the electrode is wetted by the electrolyte gel, R_s and R_d will have their minimal value since there will be a large surface area coupling the electrode to the skin. R_s will be at its minimal value because the electrolyte at the interface, consisting primarily of the gel, will be a relatively good conductor. C_d will have its greatest value since the area of contact between the electrode and the electrolyte will be large. As the electrolyte gel begins to dry, R_s will increase as the effective conductance of the electrolyte decreases as a result of decreasing contact area with the electrolytic solution. There will also be a decrease in conductivity of the electrolytic solution resulting from lower effective ion mobility due to increased concentration and salt precipitation. R_d will increase and C_d will decrease because of the decrease in effective electrode contact area. When completely dry R_d will be infinite or nearly infinite and C_d will be quite small.

38 Biopotential Electrodes

Under these circumstances, the equivalent circuit will reduce to a series combination of the new R_s and the new C_d denoted by primes as indicated below. Note that there is no longer any electrode-electrolytic solution interface to produce it. The ECG will lose its low-frequency components because C_d' and the amplifier input impedance act as a high-pass filter.

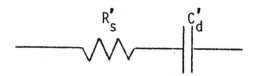

5.5   There will be a potential difference between the electrodes due to the difference in half-cell potentials for the two materials. These potentials and the associated reactions are taken from Table 5.1 and are given below:

$$Zn \rightarrow Zn^{2+} + 2e^- \qquad -0.763 \text{ V} \qquad (P\ 5.5.1)$$

$$Al \rightarrow Al^{3+} + 3e^- \qquad -1.706 \text{ V} \qquad (P.5.5.2)$$

The voltage difference between these two reactions is 0.943 V with the Al being negative, thus this potential will appear between the two wires.

5.6   Fig. 5.5 shows a minimal impedance at 100 Hz of 160 Ω for 0.25 cm². To achieve 10 Ω, make 16 times larger, or 4 cm². However the text states 10-Hz impedance of 1 cm² is less than 10 Ω and higher frequency always produces lower impedance, so 1 cm² should be less than 10 Ω at 100 Hz also. Suggest electrode of Fig. 5.16(c) with 1 cm² area.

5.7   One can determine the approximate source impedance by using either an oscilloscope or a chart recorder and a known variable resistance such as a decade box. The chart recorder and oscilloscope must have an input impedance much higher than the suspected source impedance of the electrode pair. The amplitude of the signal from the electrodes is determined with the electrodes connected directly to the indicating device. The conditions at this time should be close to those of an open circuit thereby giving an open circuit voltage for the equivalent electrical source from the electrodes. The decade resistance is now connected in parallel with this source and the indicating device, and its resistance value is varied until the signal amplitude, as seen on the indicator, has dropped to one-half of its original open circuit value. At this point, the resistance on the decade box should approximately equal the magnitude of the source resistance at the electrode provided that there is not appreciable phase shift. This technique will give a rough approximation of the magnitude of the impedance very quickly.

It is important to note that this, indeed, is only a rough approximation since loading the electrodes with a resistance equal to their source resistance will cause considerable current to flow, resulting in polarization of the electrode-electrolyte interface. This can result in a change in the effective source impedance of the electrodes, thus we are no longer measuring the same situation that we would have if the electrodes were driving a load impedance which was much larger than their source impedance. One can avoid this problem partially by not allowing the decade box to be set to a low value so as to reduce the amplitude of the signal by one-half. Instead, one could set the decade resistance so that the signal if only reduced by 10% and then, using the voltage divider equation, solve for the effective source resistance. In this case, we would not be providing as great a load to the electrode, but the calculation would magnify measurement errors.

In both of the above examples, we only are looking at equivalent source resistance of the electrodes. It is well known that when electrodes are loaded down, the signal seen from them undergoes low-frequency distortion and this indicates that there must also be a reactance associated with the electrode. The best way to determine the overall source impedance of the electrode is to place the electrode in a sensitive bridge circuit so that low excitation can be used and the resistive and reactive components of the equivalent impedance of the electrode can be determined without causing the electrode to stimulate tissue or become polarized.

5.8   2nd ed. Figure 5.8 shows changes in electrode impedance Z_e for current densities larger than 500 $\mu A/cm^2$. For a 5-V source, the current determining resistor $R = V/I = 5/0.0005 = 10$ kΩ. Then we would have to use the voltage divider equation to calculate Z_e whenever R and Z_e are comparable. It might make calculations easier to make $R = 1$ MΩ. Then we can assume a constant current of 5 μA.

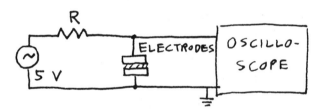

5.9   When current passes between the electrodes, a polarization potential will result such as that given by Equation 5.4. Since the current is flowing from electrode to electrolyte solution at one electrode, and in the opposite direction at the other, the sign of the three components of the polarization potential for each electrode will differ due to the current flow direction. For example, the ohmic overpotential will be in the same direction for both electrodes even though the half-cell potential for each will be of opposite polarity. Thus, the ohmic overpotential will add to the half-cell potential of one electrode and subtract from that of the other. It is reasonable to assume that the ionic concentrations at the different electrodes will be different, due to the different reactions occurring at each (one will be oxidation, the other will be reduction) and, so, this term will be different for different electrodes. Finally, the activation overpotential as described in the text is different for different directions of the current flow. Thus, all of these will add up to different net offset potentials for each electrode.

40 Biopotential Electrodes

5.10 The equivalent circuit for this situation is shown below.

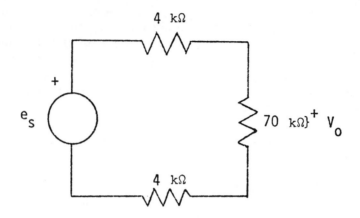

The output voltage is given by the voltage divider equation

$$V_0 = \frac{70k}{70k + 4k + 4k} e_s = 0.897 e_s \qquad \text{(P 5.10.1)}$$

Thus, the signal voltage is reduced to 89.7% of its true value.

5.11 By cleaning the silver-silver chloride electrode with steel wool, some of the silver chloride layer will no doubt be scratched off exposing elemental silver. The half-cell potential for silver is different from that of silver chloride as noted in Table 5.1 and elemental silver is more polarizable than the silver salt. Thus, the effects associated with polarization, namely increased motion artifact, noise, and low-frequency source impedance, should increase following the steel wool cleaning. In addition, the half-cell potential of the electrode will be different after the cleaning procedure. All of this acts together to give an electrode which will be more unstable and not as good for cardiac monitoring.

5.12 (a) The resistance of the metal tip of the electrode can be determined approximately from geometric considerations and the resistivity of the metal.

$$R = \rho \frac{L}{A} = 1.2 \times 10^{-5} \, \Omega\cdot cm \, \frac{3mm}{(1\mu m)^2 \pi/4} = 458 \, \Omega \qquad \text{(P 5.12.1)}$$

(b) Only the cross-section of the microelectrode tip will contact the electrolytic solution within the cell (cytosol). This surface area is

$$A = \frac{(1\mu m)^2 \pi}{4} = 7.85 \times 10^{-9} \, cm^2 \qquad \text{(P 5.12.2)}$$

The conductance of the interface will be given by the conductance per square centimeter of interface multiplied by the cross-sectional area of the tip

$$G = 7.85 \times 10^{-9} \text{ cm}^2 \times 10^{-3} \text{ S/cm}^2 = 7.85 \times 10^{-12} \text{ S} \qquad \text{(P 5.12.3)}$$

which is equivalent to a resistance of 1.27×10^{11} Ω. Clearly this is the dominant resistance for the microelectrode and the resistance due to the metal tip as calculated in part (a) is insignificant when compared to this value.

(c) The capacitance associated with the tip of this microelectrode can be calculated using Equation 5.16.

$$C = \frac{2\pi \times 1.67 \times 8.8 \times 10^{-12} \times 3 \times 10^{-3}}{\ln(1.4/1)}$$

$$= 0.823 \text{ pF} \qquad \text{(P 5.12.4)}$$

(d)

[Circuit diagram showing E_{mp} and E_{hc} sources in series with a 1.27×10^{11} Ω resistor, in parallel with a 0.823 pF capacitor, connected to load R_L.]

(e) Connecting the electrode to the amplifier places resistance R_L, shown in the figure above, across the capacitor. The ac Thevenin equivalent resistance of the electrode-amplifier combination seen from the terminals of the capacitor will then be the parallel combination of R_L and the source resistance of the electrode. Since R_L is several orders of magnitude lower than the source resistance, the Thevenin equivalent will equal approximately the resistance R_L. Thus, the circuit becomes a low-pass filter having a corner frequency given by

$$f = \frac{1}{2\pi RC} = \frac{1}{2\pi \times 10^7 \text{ Ω} \times 0.823 \text{ pF}}$$

$$= 1.92 \times 10^4 \text{ Hz} \qquad \text{(P 5.12.5)}$$

When the input impedance of the amplifier is increased to 100 MΩ the corner frequency will be one-tenth of the above value. It is seen that the increase in input impedance of the amplifier from 10 MΩ to 100 MΩ has reduced the high corner frequency of the low-pass filter to below 2000 Hz. At these frequencies, one begins to get significant distortion of depolarization and repolarization waveforms themselves, primarily present in the form of diminished rise times. Nevertheless, since the input impedance of the amplifier is much lower than the source impedance of the electrode, one does get a tenfold increase in signal amplitude from this modification. For intracellular biologic applications, it would be more desirable to use an amplifier with much higher input impedance and a negative capacitance input characteristic to both increase signal amplitude and minimize loading of the microelectrode as well as to extend the effective frequency response of the system.

5.13 The tip of the micropipet electrode can be modeled as a cylinder which is 3 μm in diameter and two millimeters long with a 0.5 μm thick layer of insulation. This layer of insulation will result in a distributed shunting capacitance along the 40 MΩ resistance due to the electrolyte in the tip. We find the capacitance by applying Equation 5.16

$$C = \frac{2\pi \varepsilon_r \varepsilon_0 L}{\ln(D/d)} = \frac{2\pi \times 1.63 \times 8.8 \times 10^{-12} \text{ F/m} \times 2 \times 10^{-3} \text{m}}{\ln(4/3)}$$

$$= 0.626 \text{ pF}$$

(P 5.13.1)

We can make an approximate equivalent circuit for this microelectrode as shown below

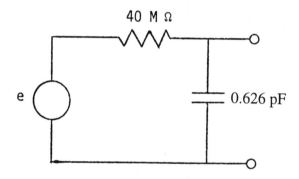

This is a low-pass filter and its corner frequency will be given by

$$f = \frac{1}{2\pi RC} = \frac{1}{2\pi \times 40 \times 10^6 \Omega \times 0.63 \times 10^{-12} \text{ F}}$$

$$= 6350 \text{ Hz}$$

(P 5.13.2)

The frequency response can be improved by reducing the effective series resistance and/or the shunt capacitance of the electrode. This is not always very easy to do with a particular microelectrode, and so one must consider what could be done with external circuit elements to produce the same effect. If a very high input impedance amplifier is used so that it provides practically no load on the electrode, we continue to have the equivalent circuit shown above. If this amplifier has its input arranged so that it appears to have a negative capacitance (see Chapter 6), if we make this capacitance just less than the capacitance of the microelectrode, this will reduce the effective capacitance seen in the low-pass filter and raise the corner frequency. As we will learn in Chapter 6, we do not want to make the negative capacitance too large since the amplifier then becomes unstable.

5.14 (a) The equivalent circuit is as follows:

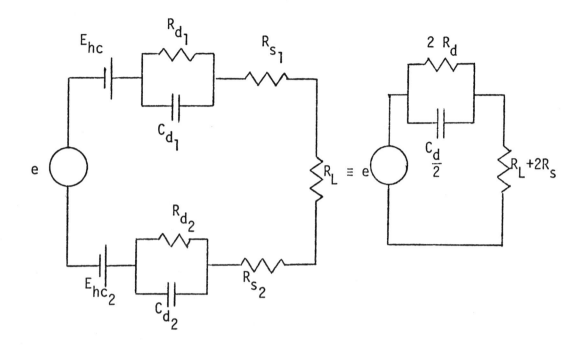

(b) If the electrodes used in this problem are identical, the equivalent circuit on the left above can be simplified to give the equivalent circuit on the right. On this circuit we see that the electrodes will tend to distort low frequencies, hence, the circuit behaves as a high-pass filter. The corner frequency of the high-pass filter will be

$$f = \frac{2(R_d)(R_L + 2R_s)}{\pi(2R_d + R_L + 2R_s)C_D} \qquad (P\ 5.14.1)$$

For values of R_L which are much greater than R_s and R_d the corner frequency is given approximately by

$$f \simeq \frac{2R_d}{\pi C_d} \qquad (P\ 5.14.2)$$

When R_L is small compared to these quantities, the corner frequency becomes

$$f \simeq \frac{2R_d R_s}{\pi(R_d + R_s)C_d} \qquad (P\ 5.14.3)$$

Thus, it is seen that the value of the input impedance of the amplifier will affect the frequency response of the electrode system. The lower the amplifier's input impedance, the higher the corner frequency of the high-pass filter becomes.

44 Biopotential Electrodes

5.15 The basic equivalent circuit for the electrode system is shown below.

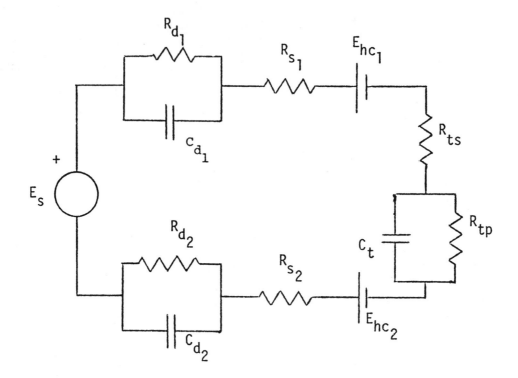

In addition to the standard equivalent circuit components for the electrodes themselves R_{ts}, C_t, and R_{tp} represent the electrical impedance of the tissue between the electrodes. If both electrodes are identical and have identical equivalent circuits, the circuit can be simplified to that shown below.

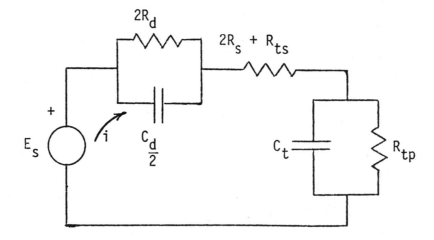

Upon applying the stimulus to this circuit, there will be two time constants of interest associated with the charging of each capacitor. Since we do not know the sizes of the capacitors and resistors, it is difficult to specify the actual waveform, but we can indicate initial and final asymptotic currents. At the beginning of the pulse, the capacitors should be discharged and appear momentarily as short circuits thereby giving a maximum current of

$$I_1 = \frac{5V}{2R_s + R_{ts}} \tag{P 5.15.1}$$

Once the capacitors are completely charged, the current becomes

$$I_2 = \frac{5V}{2R_s + R_s + R_{ts} + R_{tp}} \tag{P 5.15.2}$$

Generally, the time at which the current reaches I_2 is much greater than the duration of the pulse, thus the waveform would be truncated as indicated in the figure.

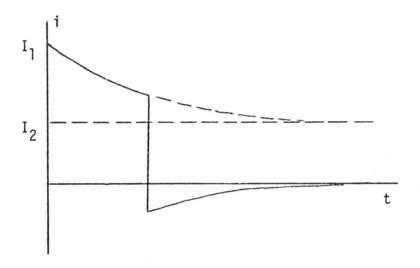

5.16 Since this new animal has the bromide ion rather than the chloride ion as its principal anion, silver-silver chloride electrodes will not work as well as they do in the more conventional animals that we know. There will still be a concentration gradient of chloride and bromide ions at the interface between the electrode and the animal's body fluids and this layer will be responsible for some of the artifact seen. A more appropriate electrode system for use with this animal would be silver-silver bromide electrodes since silver bromide is also a solid of low solubility which can be formed into an electrode using the processes described in the chapter for forming silver-silver chloride electrodes. This electrode will work in much the same way as the silver-silver chloride electrode works in animals where the principal anion is chloride, and it should give improved stability over the silver-silver chloride electrodes on this exotic animal. A limitation, however, makes this electrode not as stable as the chloride electrodes in conventional animals. This is the greater sensitivity of silver bromide to reduction in the presence of light. This tends to make the electrode less stable and can result in elemental silver being in contact with the animal and thereby making the electrode less stable electrically.

Biopotential Electrodes

5.17 (a) The circuit may be reduced to the equivalent circuit shown below.

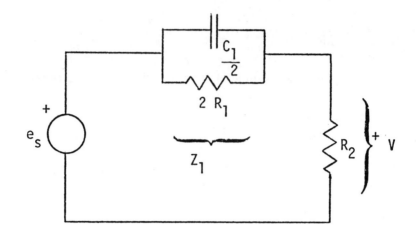

In this case, the output voltage V is

$$V = \frac{R_2}{R_2 + Z_1} e_s \qquad (P\ 5.17.1)$$

The impedance Z_1 is given by

$$Z_1 = \frac{\left(\frac{1}{SC_1/2}\right) 2R_1}{\frac{1}{SC_1/2} + 2R_1} = \frac{2R_1}{1 + S\ R_1 C_1} \quad (S = j\omega) \qquad (P\ 5.17.2)$$

which makes V

$$V = \frac{R_2}{R_2 + \frac{2R_1}{1 + S\ R_1 C_1}} e_s = \frac{R_2 (1 + S\ R_1 C_1)}{R_2 (1 + S\ R_1 C_1) + 2R_1} e_s \qquad (P\ 5.17.3)$$

$$V = \frac{R_2 + S\ R_1 R_2 C_1}{R_2 + 2R_1 + S R_1 R_2 C_1} e_s \qquad (P\ 5.17.4)$$

When e_s is sinusoidal, we can write Equation (P 5.17.4) in terms of $j\omega$

$$V = \frac{R_2 + j\omega R_1 R_2 C_1}{R_2 + 2R_1 + \omega R_1 R_2 C_1} e_s \qquad (P\ 5.17.5)$$

$$V = \frac{R_2 + j\omega R_1 R_2 C_1}{R_2 + 2R_1 - j\omega R_1 R_2 C_1} \left(\frac{R_2 + 2R_1 - j\omega R_1 R_2 C_1}{R_2 + 2R_1 - j\omega R_1 R_2 C_1}\right) \qquad (P\ 5\ 17.6)$$

$$V = \frac{R_2^2 + 2R_1R_2 + \omega^2 R_1^2 R_2^2 C_1^2}{(R_2 + 2R_1)^2 + \omega^2 R_1^2 R_2^2 C_1^2} + j\frac{\omega R_1 R_2 C_1 (2R_1)}{(R_2 + 2R_1)^2 + \omega^2 R_1^2 R_2^2 C_1^2} \quad \text{(P 5.17.7)}$$

(b) The equivalent circuit now becomes

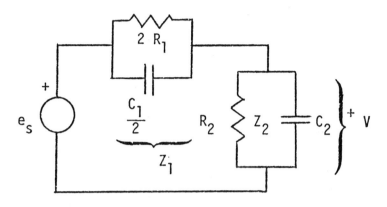

The impedance Z_2 is given by

$$Z_2 = \frac{R_2}{1 + S R_2 C_2} \quad \text{(P 5.17.8)}$$

making the output voltage

$$V = \frac{Z_2}{Z_1 + Z_2} e_s = \frac{\dfrac{R_2}{1 + S R_2 C_2}}{\dfrac{2R_1}{1 + S R_1 C_1} + \dfrac{R_2}{1 + s R_2 C_2}} e_s \quad \text{(P 5.17.9)}$$

If we require

$$R_1 C_1 = R_2 C_2 = \tau \quad \text{(P 5.17.10)}$$

then Equation (P 5.17.9) becomes

$$V = \frac{\dfrac{R_2}{1 + S\tau}}{\dfrac{2R_1}{1 + S\tau} + \dfrac{R_2}{1 + S\tau}} e_s = \frac{R_2}{2R_1 + R_2} e_s$$

which is independent of frequency

(c) Since the gain of the amplifier in the problem is A, the output voltage from it when the conditions of Equation (P 5.17.10) are met will be

5.18 The low corner frequency is $F_c = 1/(2\pi RC) = 1/(2\pi \cdot 20 \text{ k}\Omega \cdot 100 \text{ nF}) = 80$ Hz.
The high corner frequency is $F_c = 1/(2\pi RC) = 1/(2\pi \cdot 20 \text{ k}\Omega \| 300 \text{ }\Omega \cdot 100 \text{ nF}) = 5380$ Hz.

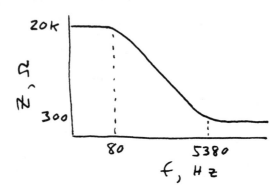

5.19 The 600 Ω is the tissue impedance plus the electrode/electrolyte high-frequency interface impedance. The 19400 Ω is the electrode/electrolyte low-frequency interface impedance. The 200 nF is the electrode/electrolyte interface capacitance. The 233 mV is the electrode/electrolyte polarization voltage.

Chapter 6
Biopotential Amplifiers

Michael R. Neuman

6.1 From (6.1), assume $|\bar{M}| = 1$, $\theta = 0°$ for $\bar{a}_I$.

$$\begin{aligned}
\text{Sum} &= \bar{M} \cdot \bar{a}_I + \bar{M} \cdot \bar{a}_{II} + \bar{M} \cdot \bar{a}_{III} \\
&= \cos\theta + \cos(\theta - 60°) + \cos(\theta - 120°) \\
d\text{Sum}/d\theta &= 0 = -\sin\theta - \sin(\theta - 60°) - \sin(\theta - 120°) \\
&= -\sin\theta - \sin\theta\cos 60° + \cos\theta\sin 60° - \sin\theta\cos 120° + \cos\theta\sin 120° \\
&= -\sin\theta - 0.5\sin\theta + 0.86\cos\theta + 0.5\sin\theta + 0.86\cos\theta
\end{aligned}$$

$\sin\theta/\cos\theta = \tan\theta = 1.73$

$\theta = 60°$ (in the direction of $\bar{a}_{II}$)

6.2 From (6.1), assume $|\bar{M}| = 1$, $\theta = 0°$ for $\bar{a}_I$.

$\bar{M} \cdot \bar{a}_{II} = \bar{M} \cdot \bar{a}_{III}$.

$\cos(\theta - 60°) + \cos(\theta - 120°)$

$\cos\theta\cos 60° + \sin\theta\sin 60° = \cos\theta\cos 120° + \sin\theta\sin 120°$

$0.5\cos\theta + 0.86\sin\theta = -0.5\cos\theta + 0.86\sin\theta$

$\cos\theta = 0$; $\theta = 90°, 270°$

(in the direction of $\bar{a}_{aVF}$ or $-\bar{a}_{aVF}$)

The R wave in lead I would equal zero because $\bar{a}_I$ is orthogonal to $\bar{a}_{aVF}$. In actual practice the QRS complex would appear to be biphasic because the cardiac vector changes magnitude and direction during the QRS complex and will only be in the $\bar{a}_{aVF}$ direction during the peak of the R wave.

6.3 Use Eq. (6.2). $I - II + III = 0 = I - 1.0 + 0.5$, from which $I = 0.5$.

6.4

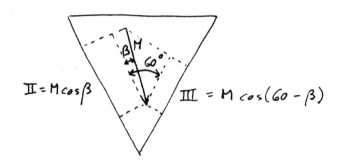

$III = M\cos(60° − ß) = M(\cos 60°\cos ß + \sin 60°\sin ß) = M(0.5\cos ß + 0.86\sin ß) =$
$M[0.5(II/M) + 0.86(1 − \cos^2 ß)^{0.5}] = M[II/2M + 0.86(1 − II^2/M^2)^{0.5}]$
$III/M − II/2M = 0.86(1 − II^2/M^2)^{0.5}$
$2III − II = 1.73M(1 − II^2/M^2)^{0.5}$
$(4III^2 − 4II·III + II^2) = 3M^2(M^2 − II^2)/M^2$
$4III^2 − 4II·III + II^2 = 3M^2 − 3II^2$
$4III^2 − 4II·III + 4II^2 = 3M^2$
$(4/3)(III^2 − II·III + II^2) = M^2$
$1.15(III^2 − II·III + II^2)^{0.5} = M$

(implement with computer)

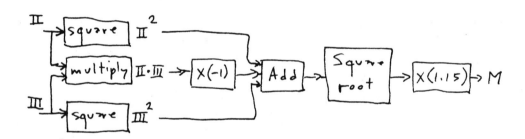

6.5   The Wilson central terminal in Fig. 6.4 has a total resistance for one electrode, when all others are connected to ground of R + R‖R = R + R/2 = 3R/2 = 2.5 MΩ (from Table 6.1). Hence R = 1.7 MΩ for VF. R/3 = 0.57 MΩ resistor in series with LL balances Thevenin source impedance to minimize interference. For aVF in Fig. 6.5(c), 2R = 2.5 MΩ so R = 1.25 MΩ. R/2 = 0.625 MΩ resistor in series with LL balances Thevenin source impedance to minimize interference.

6.6   From Fig. 6.5(d), aVF is directed toward the feet. Invert the system of Fig. 6.5(c) by placing LL electrode on neck and LA and RA electrodes on hips. Table 6.1 shows input impedance should be 2.5 MΩ for any lead with all others grounded. Therefore let R = 1.25 MΩ to achieve 2R = 2.5 MΩ. R/2 = 0.625 MΩ resistor in series with LL balances Thevenin source impedance to minimize interference.

6.7 The switching function can be achieved using a three-circuit six-position rotary selector switch as shown below.

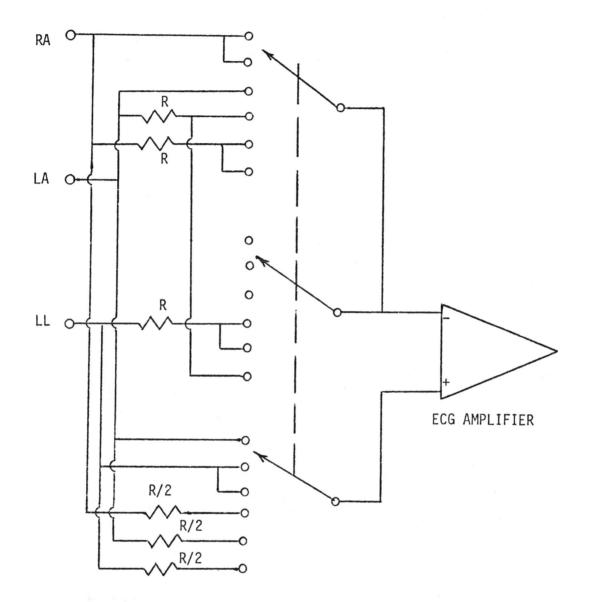

The positions of the selector switch from top to bottom are:
I, II, III, aVR, aVL and aVF.

6.8 Each pair of electrodes connected in the Wilson central terminal configuration sees a load resistance of 2R. Thus, if we make R small, the effect will be to apply a significant load to the electrodes and in doing so, distort the electrocardiogram. If R is small, variations in the source impedances of the different electrodes will cause the effective R for each electrode to be different. This will then distort the Wilson central terminal and not make it "truly central." If the values of R are, on the other hand, too large, they will present a significantly large source resistance for the

52 Biopotential Amplifiers

Wilson central terminal. This source resistance must be balanced by an equivalent source resistance in the other lead. The effect of large source resistances is to increase circuit noise, interference and offset voltages.

6.9 Table 6.1 states that the input impedance between an electrode terminal and ground should be no less than 2.5 MΩ. All other electrodes during the measurement should be grounded. With the selector switch on V leads, the equivalent circuit would be:

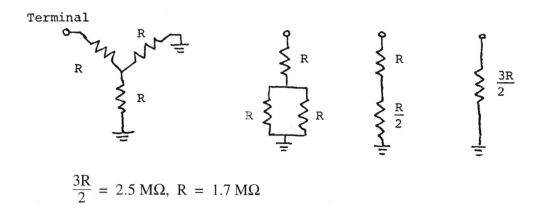

$$\frac{3R}{2} = 2.5 \text{ M}\Omega, \quad R = 1.7 \text{ M}\Omega$$

6.10 The circuit in Figure 6.9 can be reduced to the equivalent circuit shown below.

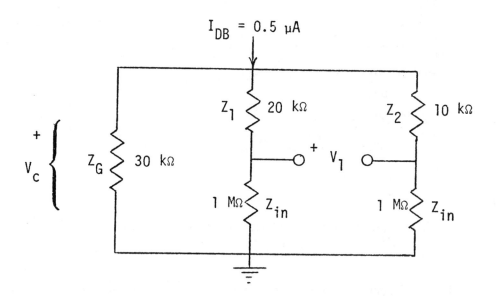

The common mode voltage of the body above ground potential can be determined by multiplying the displacement current by the net parallel resistance of the body and amplifier.

$$V_c = 0.5 \,\mu A \left(\frac{1}{\frac{1}{30\text{ K}} + \frac{1}{1.02\text{ M}} + \frac{1}{1.01\text{ M}}} \right) = 14.1 \text{ mV} \tag{P6.10.1}$$

Now we can determine the amount of interference voltage on V_1 by using the voltage divider equation.

$$V_0 = \left(\frac{1\text{ M}}{1.02\text{ M}} - \frac{1\text{ M}}{1.01\text{ M}} \right) 14.1 \text{ mV} = -0.137 \text{ mV} \tag{P6.10.2}$$

This means that the power line interference observed on the electrocardiogram will be 137 μV referred to the input of the electrocardiograph

6.11 Let us add a driven right leg amplifier to the basic ECG amplifier shown in Figure 6.16. The circuit for this addition is shown below. Let us first find the value of the output resistor R_o. The requirement that no more than 5 μA flows when the amplifier is saturated at 12 V, gives us enough conditions to determine

$$R_0 = \frac{12 \text{ V}}{5 \,\mu A} = 2.4 \text{ M}\Omega \tag{P6.11.1}$$

Since one cannot obtain standard 10% resistors at 2.4 MΩ, we can go to the next higher value which would be 2.7 MΩ to ensure that the current will be less than 5 μA.

Then $v_0 = I_D R_0 = (1 \,\mu A)(2.7 \text{ M}\Omega) = 2.7 \text{ V}$

Equation (E6.10) yields:

$R_f = -v_0 R_a / 2v_C = 2.7(10 \text{ k}\Omega)/2(0.002) = 6.75 \text{ M}\Omega$

Choose the next higher standard value, 6.8 MΩ

54 Biopotential Amplifiers

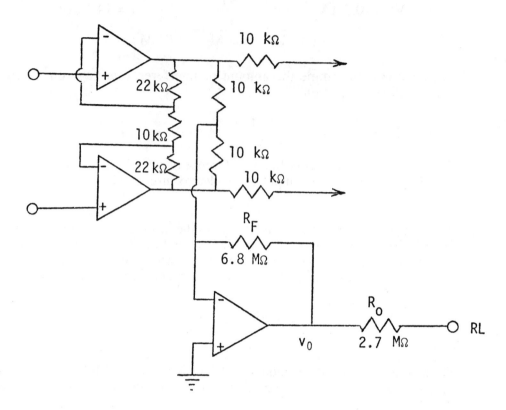

6.12 By shorting out the resistance noted as R/2 in Figure 6.5a, the engineer has unbalanced the input to the amplifier used to amplify the signal from this lead. The source impedance seen by the positive and negative inputs of the amplifier are no longer equal, and input bias current of the amplifier can result in an offset voltage at the input for this reason. In addition, the no longer balanced source for this amplifier will reduce the common mode rejection ratio of the amplifier and, therefore, is likely to contribute more noise to the resulting electrocardiogram.

6.13 From Figure 6.14 we see that frequency components of the ECG and EMG overlap. Thus, it is not possible to completely separate the two signals; however, by using a high-pass filter we can separate most of the EMG signal from the composite signal, and by using a low-pass filter, we can separate out much of the ECG. By choosing a corner frequency of 50 Hz, we can obtain fairly good separation. The circuit to be use is shown below, and the design is based upon a procedure described by J. G. Graeme, G. E. Tobey and L. P. Huelsman; Operational Amplifiers, Design and Applications, McGraw-Hill, New York, 1971, in Section 8.3.1.

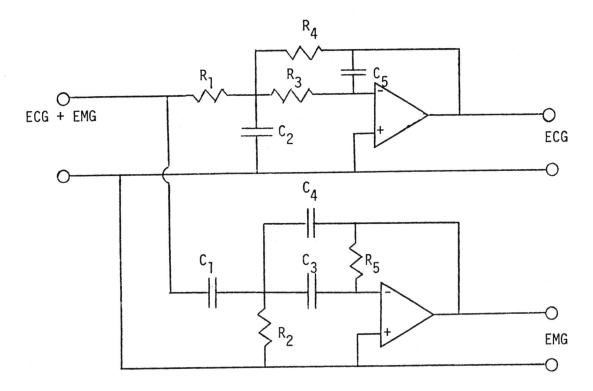

Let us first examine the low-pass filter for obtaining the ECG. Based on the reference given above, the gain of this active filter at frequencies below the corner frequency will be

$$A_L = \frac{R_4}{R_1} \tag{P 6.13.1}$$

The damping ratio is given by

$$\xi = \sqrt{\frac{C_5}{C_2}} \left(\sqrt{\frac{R_3}{R_4}} + \sqrt{\frac{R_4}{R_3}} + \frac{\sqrt{R_3 R_4}}{R_1} \right) \tag{P 6.13.2}$$

and the corner frequency is given by

$$f_0 = \frac{1}{2\pi} \sqrt{\frac{1}{R_3 R_4 C_2 C_5}} \tag{P 6.13.3}$$

A gain of 10 would be convenient for this circuit at low frequencies and a damping ratio of 1 will give us a good low-pass filter function without any peaking. Let us set R_3 and R_4 to be 100 kΩ. Then equation (P 6.13.1) requires that R_1 be 10 kΩ. We can now solve equation (P 6.13.2) for the capacitor ratio.

$$1 = \sqrt{\frac{C_5}{C_2}} \, (1 + 1 + 10) \tag{P 6.13.4}$$

$$\frac{C_5}{C_2} = 6.94 \times 10^{-3} \tag{P 6.13.5}$$

We can now use equation (P 6.13.3) and (P 6.13.5) to obtain the capacitor values.

$$(50 \times 2\pi)^2 = \frac{1}{(10^5)^2 \times 6.94 \times 10^{-3} C_2^2} \tag{P 6.13.6}$$

$$C_2 = 6.77 \times 10^{-6} \text{ F} \tag{P 6.13.7}$$

$$C_5 = 4.70 \times 10^{-8} \text{ F} \tag{P 6.13.8}$$

The design equations for the high-pass filter, taken from the same reference, are given below.

$$A_N = \frac{C_1}{C_4} \tag{P 6.13.9}$$

$$f_0 = \frac{1}{2\pi} \sqrt{\frac{1}{R_2 R_5 C_3 C_4}} \tag{P 6.13.10}$$

$$\xi = \sqrt{\frac{R_2}{R_5}} \frac{C_1}{\sqrt{C_3 C_4}} + \sqrt{\frac{C_3}{C_4}} + \sqrt{\frac{C_4}{C_3}} \tag{P 6.13.11}$$

Again we shall choose the high-frequency gain of the amplifier to be 10, the corner frequency to be 50 Hz, and the damping ratio to be 1. Let us set the capacitors C_3 and C_4 to be equal and assign a value of 1.0 μF to them. Capacitor C_1 can then be determined from equation (P 6.13.9).

$$C_1 = 10 \, (1.0) = 10 \, \mu\text{F} \tag{P 6.13.12}$$

We can now calculate the ratio of R_2 to R_5 from equation (P 6.13.11).

$$1 = \sqrt{\frac{R_2}{R_5}} \left(\frac{1}{0.1} + 1 + 1 \right) \tag{P 6.13.13}$$

$$\frac{R_2}{R_5} = 6.94 \times 10^{-3} \tag{P 6.13.14}$$

Using equation (P 6.13.10), it is now possible to calculate the resistance values.

$$(50 \times 2\pi)^2 = \frac{1}{6.94 \times 10^{-3} \, R_5^2 (10^{-12})} \tag{P 6.13.15}$$

$$R_5 = 6.77 \times 10^5 \, \Omega \qquad (P\ 6.13.16)$$

$$R_2 = 4.7 \times 10^3 \, \Omega \qquad (P\ 6.13.17)$$

Since some of the values calculated are not standard component values, they have been slightly modified to standard values when shown on the circuit diagram. It is important to point out that in this case one would really be better off using 1% tolerance components with values closer to the calculated values so that the characteristics of the filters more closely approximate those which are desired.

6.14 If the 60-Hz interference is due to electrostatic field pickup, it could either be a result of greater coupling between the power line and one of the ECG electrodes or lead wires, or that there is equal pickup on the patient or on both lead wires giving a common mode signal which was not completely eliminated due to the poor common mode rejection ratio of the amplifier. In the latter case, grounding the patient, if he is not already grounded, through the right leg, should reduce the interference since much of the displacement current will be returned to ground through the ground electrode. This reduces the common mode voltage and, hence, common mode signal. The common mode signal can also be reduced by shielding the patient and the lead wires from the power line by placing them in a grounded metal enclosure. In this case, the displacement current itself will be reduced. If only the lead wires are to be shielded, this can easily be done using shielded cables in which a wire braid, which is grounded, surrounds the insulated conductor. If, in either case, the amplitude of the 60-Hz interference on the electrocardiogram is diminished, this strongly suggests that the source of interference was from an electric field.

Voltages induced in the ECG system from a magnetic field will increase as the area between the lead wires connected to the differential input of the amplifier is increased. In other words, as the number of magnetic flux lines linking the circuit loop formed by the amplifier input, the two lead wires and the patient increases, the voltage induced in this loop will also increase. Thus, one can twist the lead wires together over as much of their run as is possible and, thereby, diminish the effective area of the loop and, hence, the number of magnetic flux lines linking it. This should result in a reduction of the 60-Hz interference if, indeed, its source is a magnetic field.

In both cases, relocating the patient can alter the amount of 60-Hz interference seen since this will, no doubt, change the coupling capacitances between the patient and the hot side of the power lines. One can often reduce the amount of magnetic flux linking the lead wires by turning off electrical apparatus in the vicinity of the patient. This is especially true for apparatus which draws heavy current such as motors, air conditioners and electric heaters. In rooms which are lit by fluorescent lights, these lights can produce high-frequency electrostatic fields which are modulated at the power line frequency and these can be detected by the input stages of an electrocardiograph. In some cases, the nonlinear current–voltage characteristic of the electrode–electrolyte interface can also detect such signals. The result is that the modulation signal, in this case the 60-Hz interference, appears at the input to the amplifier. This source of interference can be eliminated by turning off the fluorescent lights. X-ray and diathermy apparatus in the vicinity of the ECG system can also produce interference of this type.

6.15 Use Fig. 1.6(a) in series with each differential amplifier input. To achieve 0.001, use Eq. (1.23) from which $1000 = (1 + \omega^2\tau^2)^{0.5}$. $1000 \approx \omega\tau$. For $\omega = 2\pi 10^6$, $\tau = 160$ μs. Choose $R = 100$ kΩ, then $C = 1600$ pF.

60-Hz impedance is $1/(2\pi fC) = 1/(2\pi 60 \cdot 1600 \text{ pF}) = 1.7$ MΩ. Use Eq. (6.9), $\Delta v = 10$ mV$[(50$ kΩ $- 40$ kΩ$)/1.7$ MΩ$] = 59$ μV.

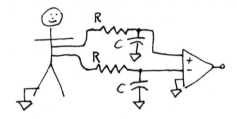

6.16 Use Eq. (6.9). $\Delta v = 10$ μV $= 10$ mV$[(100$ kΩ $- 80$ kΩ$)/X_C]$. $X_C = 20$ MΩ $= 1/(2\pi 60 C)$. $C = 133$ pF.

6.17 Let us consider the equivalent circuit for the case where V_e, the input voltage to the protection circuit, is positive. Only one of the diodes will be conducting, and we can neglect the other diode. The equivalent circuit then will be that shown below. When V_e is of the opposite polarity, the equivalent circuit is modified only by reversing the polarity of the 600 mV voltage source.

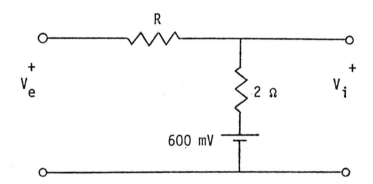

Solving for the input voltage to the preamplifier V_i, we get

$$V_i = \frac{2\Omega}{R + 2\Omega} 500 \text{ V} + 600 \text{ mV} = 800 \text{ mV} \tag{P6.17.1}$$

With this set to 800 mV, the maximum input voltage to the preamplifier, we can solve for R.

$$R = 4998 \text{ Ω} \tag{P6.17.2}$$

6.18 Use Eq. (E6.12). 10 μV = 100 kΩ(500 nA)/(1 + 2R_f/R_a). (1 + 2R_f/R_a) = 5000. Choose R_f = 5 MΩ. Then R_a = 2 kΩ.

6.19 Use Eq. (6.14). 10 pF = (A_v − 1)20 pF. A_v = 1.5. Use Fig. 3.4(b) noninverting amplifier with R_f = 5 kΩ and R_i = 10 kΩ.

6.20 A simple design is shown in the block diagram. The first part of the design consists of getting a pulse which indicates the beginning of a recording of a particular lead of the electrocardiogram. This beginning can either be a result of starting the chart on the electrocardiograph or the chart can be running and a new lead is selected by changing the lead selector switch. Either event should cause the 1 mV standardizing pulse to appear at the beginning of the recording strip. By placing an extra contact on the start switch and a contact which is broken and reestablished each time the lead selector switch is changed, one can produce a pulse indicating the desired events. Since it is not known which event initiates a new recording, both switches should be connected to an OR gate the output of which triggers a 2 s delay circuit. This circuit should be a monostable multivibrator which is triggered on the rising edge of the pulse at its input so that it only will be triggered if the machine is started or if the lead selector switch has been changed. Following the 2 s delay, the second monostable multivibrator produces a 0.2 s pulse which turns on the analog switch adding the 1 mV standardizing signal to the input to the electrocardiograph to produce the appropriate standardizing pulse. The 2-s delay is required since there is often a transient associated with the initial starting of a recording, and if the standardization pulse is added to this transient immediately at the beginning of the recording, it might deflect the pen off scale. By waiting until a large fraction of the transient has had a chance to die out, this problem can be avoided.

60 Biopotential Amplifiers

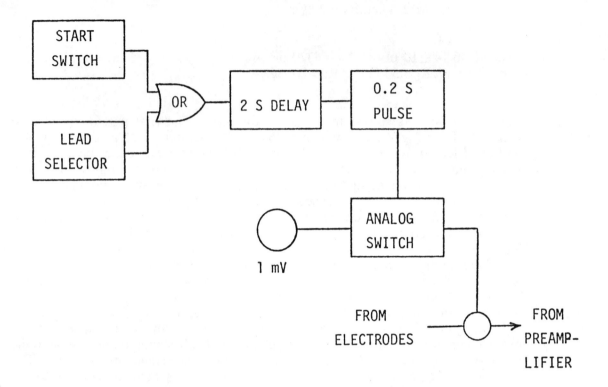

6.21 It will look like Fig. 6.7 until the switch closure. After switch closure it will take even longer to recover than Fig. 6.7(c) because worst case electrode offset potential of 300 mV times gain of 25 = 7.5 V dc at left end of C. After switch closure, right end of C will also be at 7.5 V. v_o saturates at 10 V, when right end of C = 10 V/32 = 0.31 V. Thus we must wait $e^{-t/\tau} = e^{-t/3.3 \text{ s}} = 0.31/10 = 0.031$. t = 3.47·3.3 = 11.5 s for the trace to reappear.

6.22 Choose largest practical nonpolar capacitor 10 µF. Then for $\tau = RC = 3.3$ s, R_i = 330 kΩ. R_f = 10 MΩ yields a first stage ac gain of 33 and dc gain of 1. Second stage gain of 30 ensures no saturation even when input offset voltage = 300 mV. Switch S shortens the RC time constant to yield rapid recovery.

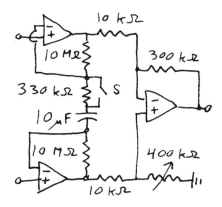

6.23 This problem involves the design of an isolation circuit to isolate a preamplifier from the output circuit. there are several ways this isolation circuit can be designed as pointed out in Chapter 14. We will choose to use a carrier signal and an isolation transformer. The isolated preamplifier could consist of the ECG amplifier illustrated in Figure 6.16. The output of this amplifier should be nominally 800 mV peak which is used to drive a voltage to frequency converter. These circuits are available in monolithic integrated circuit form. An external capacitor is placed on the voltage-to-frequency converter so that it has a nominal operating frequency of 2 kHz when there is no input. The peak signal from the amplifier should then be capable of providing approximately 30 to 40% frequency deviation to effectively use this frequency modulation scheme. The output from the voltage-to-frequency converter should be connected to the primary of a high-frequency isolation transformer with a turns ratio of 1:1. This transformer should be constructed so that there is less than 0.5 pF coupling capacitance between its primary and secondary and it should have a relatively flat frequency response from 1–3 kHz. All circuitry connected to the primary side of the isolation transformer should be battery powered and isolated from ground. Thus, in the circuit of Figure 6.22, all ground connections should go to battery common rather than to ground.

The secondary of the isolation transformer should drive a frequency-to-voltage converter which has been adjusted to have zero output at a frequency of 2 kHz. The output from this converter must then be passed through a low-pass filter with a corner frequency of 150 Hz so that artifact resulting from the frequency-modulated carrier will be removed from the output.

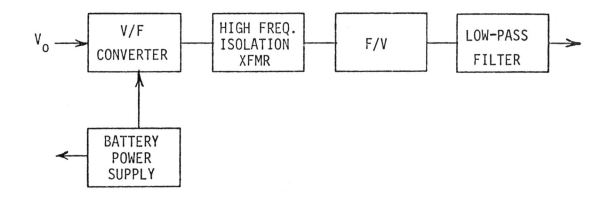

62 Biopotential Amplifiers

6.24 Design a differentiator for use for all frequencies of the ECG up to 100 Hz. Choose R = 10 kΩ. Then C = $1/(2\pi 100 \cdot 10\text{ k}\Omega)$ = 160 nF. This outputs a high voltage for the R wave and a low voltage for the T wave. The output comparator is adjusted to flip only for R waves.

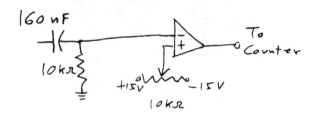

6.25 Use 2nd ed. Fig. 3.7 to design a window comparator that has a high output for $|v_i|$ > 10 V. Attach to v_o of Fig. 6.16, the voltage in the circuit that is first to saturate. Use high output to trigger a 100 ms one shot that closes S_1 of Fig. 6.16. The one shot is required to make sure v_o returns all the way to 0 V and does not stop at 9 V.

6.26 Eq. (6.15) shows R is proportional to T_R. Therefore R should increase linearly with T_R. A slow heart rate f yields a high number of clock pulses n as in the last row. A parallel arrangement is not correct because then conductance G is proportional to T_R = 1/R, which is nonlinear.

n	f
000	∞
001	1
010	1/2
011	1/3
100	1/4
101	1/5
110	1/6
111	1/7

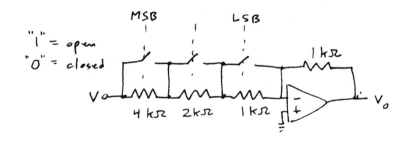

6.27 See solution for problem 6.26 or alternative solution below.

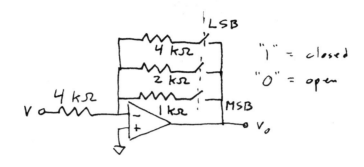

6.28 A PVC occurs early. The algorithm should test that the $R-R_{t-1}$ interval < 0.8 AR–R_{t-2}. (AR–R = average of 10). A PVC is followed by a compensatory pause. The algorithm should test that $R-R_{t-1} + R-R_t \approx 2(AR-R_{t-2})$. A PVC is wide. The algorithm should test that width $W > 1.3$AW. All three tests should be positive.

6.29 At the onset of cardiac arrest, the heart stops beating immediately, and no further QRS complexes will be seen on the electrocardiogram. Since the beat-to-beat cardiotachometer gives an output voltage which is proportional to the reciprocal of the R–R interval from the most recent to the next previous beat, its output will reflect the R–R interval preceding cardiac arrest. This output will be maintained at the cardiotachometer until the next QRS complex comes along. In the case of a cardiac arrest, this next beat may never occur. Thus, the beat-to-beat cardiotachometer indicates a heart rate inconsistent with cardiac arrest.

On the other hand, the averaging cardiotachometer will give a trace which, although it does not have as great a sensitivity to the beat-to-beat variation in heart rate as does the instantaneous cardiotachometer, will respond to a cardiac arrest. Since each QRS complex charges a capacitor which is then slowly discharged through a shunting resistor, when the QRS complexes cease, no charge will be added to the capacitor but it will continue to be discharged. Thus the output voltage from the averaging cardiotachometer will decay exponentially with a time constant of several seconds. After approximately five time constants following the cardiac arrest, this output will be essentially zero.

6.30 This circuit design is added to the basic circuit shown in Figure 6.16. You will note that the first stage of the circuit shown below is identical in configuration with the last stage of the circuit in Figure 6.16. It is repeated here because since it is a band pass filter, its characteristic frequencies must be changed to accommodate EMG rather than ECG signals. Thus, it is repeated here with different component values than it had in Figure 6.16.

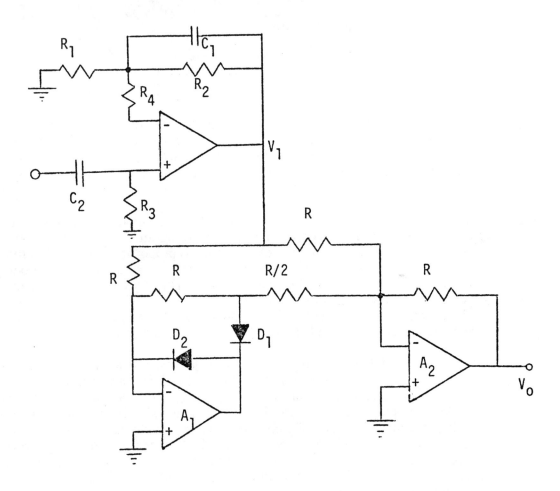

A high-pass filter is connected to the positive going input of the operational amplifier of this stage. Its corner frequency is given by

$$f_n = \frac{1}{2\pi R_3 C_2} \tag{P6.30.1}$$

From Figure 6.14 we see that the approximate range of EMG signals is from 25 Hz to about 2 kHz. Let us choose 25 Hz as the low corner frequency of the amplifier and, further, let us set R3 to be 100 kΩ. C2 is then found to be

$$C_2 = \frac{1}{2\pi \times 25 \times 10^5} = 6.37 \times 10^{-8} = 0.0637 \ \mu F \tag{P6.30.2}$$

Similarly, for the low-pass filter in the feedback circuit, the corner frequency is given by

$$f_L = \frac{1}{2\pi R_2 C_1} \tag{P6.30.3}$$

$$C_1 = \frac{1}{2\pi \times 2 \times 10^3 \times 1.5 \times 10^4} = 5.31 \times 10^{-9}$$

$$= 0.00531 \; \mu F$$

(P6.31.4)

To keep the amplifier balanced, we must set R_4 at 100 kΩ, and to maintain the output stage gain at 32, R_1 must be 470 Ω. These component values are shown on the schematic diagram as the closest standard 10% tolerance values. This is acceptable to this filter since the exact corner frequencies are not critical.

The remaining two stages constitute the absolute value circuit. It is necessary to have the preamplifier before these stages so that a signal of sufficient amplitude is obtained to minimize crossover distortion resulting from the offset voltages of the diodes in the circuit. For positive going inputs, amplifier A_1 will have a negative going output which turns on diode D_1. This makes this amplifier a unity gain inverter. The output of this inverter is summed with the input signal V_1 by the last stage. For V_1 positive, the output of the circuit is given by

$$V_0 = -\frac{10\,k}{5\,k}(-V_1) - \frac{10\,k}{10\,k}V_1 = V_1 \tag{P6.30.5}$$

When V_1 is negative, the output of amplifier A_1 will try to go positive and, hence, will turn on diode D_2 while turning off diode D_1. The feedback current through diode D_2 will then tend to maintain the inverting input of A_1 at its virtual ground level. The only input to amplifier A_2 will thus be the signal V_1. A_2 is now a unity gain converter so V_0 is not $-V_1$.

6.31 When a signal averager is used, the random noise seen on a signal will diminish by a factor of $1/\sqrt{N}$ where N is the number of repeated signals which are averaged. Thus, for a 10 to 1 improvement in the signal-to-noise ratio, N would have to be 100. For a 100 to 1 improvement in the signal-to-noise ratio, N would have to be $(100)2 = 10,000$, an impractical number.

66 Biopotential Amplifiers

6.32 A simple carrier drop-out indicator is shown below.

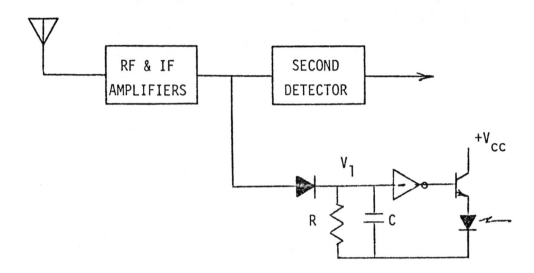

The receiver is represented by a block for the RF and IF amplifiers and another block for the second detector. Between these two blocks the IF signal is picked off and passed through a diode detector. The R–C filter should have a time constant which is long with respect to the IF carrier period but not sufficiently long to significantly delay the lighting of the indicator. Thus, when the carrier is present, voltage V_1 will be greater than zero. This voltage is inverted and used to drive a light indicating diode which is illuminated in the absence of a carrier.

6.33 A simple first approximation to orthogonal leads in the frontal plane would be to look at the standard frontal plane leads and their vectors to see if any are already orthogonal. You will find that lead I and lead aVF are indeed orthogonal using the notation of Eindhoven's triangle. In reality, however, due to the asymmetry of the chest, the position of the heart in the chest, the conductivity of the blood, and other less important factors, the Eindhoven triangle is not always a good representation of the lead vectors from the standard limb leads. To test whether these two leads are indeed orthogonal, one must make a measurement.

A simple test of orthogonality is to generate a dipole moment in the direction of one lead vector and try to detect it with the other lead. If the two leads are truly orthogonal, it should not be possible to detect this dipole moment with the second lead. The dipole moment can be generated along one lead vector by connecting a low amplitude sinusoidal voltage to the terminals of that lead. This will set up an oscillating dipole along the lead vector for that lead due to the current distribution in the volume conductor. The second lead can then be connected to the input of a sensitive oscilloscope and one can look for the signal due to the stimulus applied to the first lead. If no signal or only very small signal is seen, the leads may be considered to be orthogonal.

6.34 The ECG shown demonstrates low-frequency distortion due to diminished amplifier gain at low frequencies. This is often the result of either a defective amplifier design wherein the time constant of coupling capacitor networks between stages is not sufficiently long, or, the ECG electrodes themselves are heavily loaded because the input impedance of the amplifier is not high with respect to the source impedance of the electrodes. The solution to this problem consists of either directly coupling the stages of the amplifier in the case where the problem is attributed to the coupling circuit, or, where the problem is attributed to loading, the input impedance of the amplifier should be significantly increased.

6.35

68 Biopotential Amplifiers

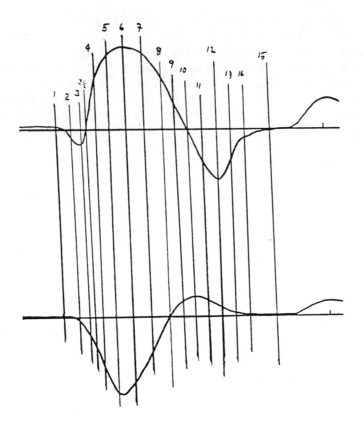

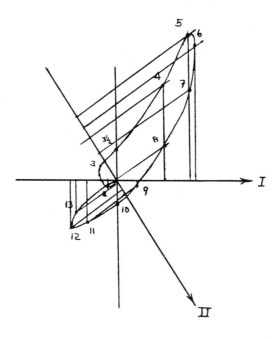

Chapter 7
Blood Pressure and Sound

Robert A. Peura

7.1 The simplest method is the transient step response since a minimum of specialized equipment is necessary. The interpretation of the response is straightforward provided that the frequency response of the recording instrument is adequate and the system behaves as a simple second-order model. Higher order responses become more difficult to interpret. The sinusoidal frequency response method is potentially more accurate but requires more specialized equipment. It requires a sinusoidal pressure-generator test system and a reference pressure-sensor system. Accurate measurements of the catheter-sensor system characteristics may be found by determining the amplitude and phase as a function of frequency, without the constraint of the second-order system model required in the transient-response case.

7.2 (a) The damping ratio ζ, can be found from

$$\zeta = \frac{\Lambda}{\sqrt{4\pi^2 + \Lambda^2}} = 0.20 \tag{1.38}$$

where $\Lambda = \ln 5 - \ln 1.4 = 1.27$

(b) The frequency of oscillation of the damped sine wave is $\omega_d = 2\pi/T$ where T = period = 0.27 ms

$$\omega_n = \omega_d/\sqrt{1-\zeta^2} = 2\pi/T \sqrt{1-\zeta^2} = 239 \text{ r/s}$$

(c) Frequency response curve

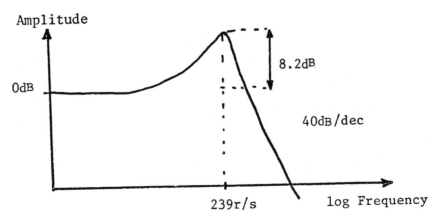

69

70 Blood Pressure and Sound

7.3 From P 7.3, $\zeta = 0.2$ for water at 20°C. Use the relationships for η, (7.6) and (7.7) which are found in the text. It is assumed that the compliance and density of blood remain the same with a temperature increase from 20°C to 37°C. Thus, f_n given by (7.6) does not change in this problem. However ζ increases by a factor of 2.8 due to an increase in η of 2.8.

$$\zeta = \frac{4\eta}{r^3}\left(\frac{L(\Delta V/\Delta P)}{\pi \rho}\right)^{1/2} \tag{7.7}$$

Therefore $\zeta_{blood @ 37°} = 0.56$

The new frequency response curve is given below.

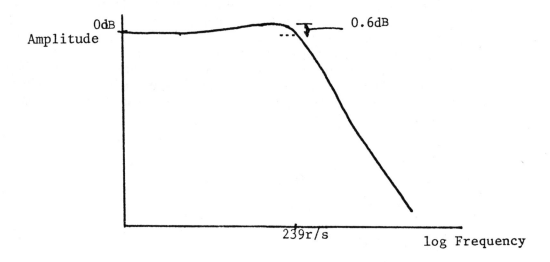

From this problem it is noted that a sensor system becomes more damped when filled with the blood due to the increase in viscosity of blood.

7.4 This system is the same as that of Example 7.1. Thus we have $f_n = 91$ Hz and $\zeta = 0.033$ for the system without a leak. The system with a leak may be modeled as shown below.

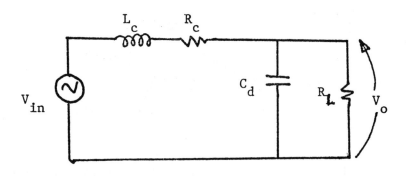

Blood Pressure and Sound

This circuit is similar to the circuit in Fig. 7.8 with R_L connected in parallel to C_d. R_L represents the pinhole leak at the junction of the catheter and sensor. We need to find new relationships for f_n and ζ for this circuit.

$$V_{0\,(j\omega)}/V_{i\,(j\omega)} = \frac{(1/C_d L_c)}{-\omega^2 + j\omega\left(\frac{R_L R_c C_d + L_c}{R_L C_d L_c}\right) + \frac{R_L + R_c}{R_L C_d L_c}}$$

The standard form for a second order system is

$$V_0(j\omega)/V_i(j\omega) = \frac{K\omega_n^2}{-\omega^2 + 2\zeta\omega_n(j\omega) + \omega_n^2}$$

Thus $\omega_n = \sqrt{(R_L + R_c)/R_L C_d L_c}$

and $\zeta = (R_L R_c C_d)/(2 R_L C_d L_c)\sqrt{R_L C_d L_c/R_L + R_c}$

The values for R_L, R_c, C_d and L_c must be determined

$C_d = 1/E_d = 2.04 \times 10^{-15}$ m^5/N

$L_c = \rho L/\pi r^2 = (1\times 10^3)(1)/\pi (0.046\times 10^{-2})^2 = 1.5 \times 10^9$ Pa•s^2/m^3

$R_c = 8\eta L/\pi r^4 = 8(0.001)(1)/\pi (0.046\times 10^{-2})^4 = 4.69 \times 10^{10}$ Pa•s/m^3

R_L = Pressure/Flow = 13.3 kPa/ (0.4 ml/min)(1 min/60 s)
$\times (1 \times 10^{-3}$m^3/1000 ml)
= 2×10^{12} Pa•s/m^3

Thus for the system with a leak

$\omega_n = 580$ r/s or $f_n = 92.4$ Hz

and

$\zeta = 0.24$

For reference the parameters for the system without a leak

$f_n = 91$ Hz
$\zeta = 0.033$

72 Blood Pressure and Sound

The frequency response curves for the system with and without the leak are given below.

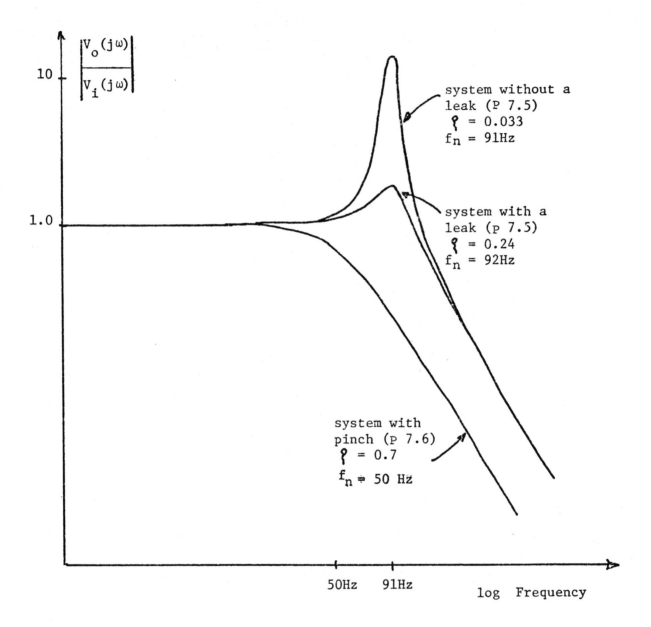

7.5 From Eq. (7.7), $r^3/r_0^3 = \zeta/\zeta_0$.

$r^3 = (1.0/0.033)(0.46 \text{ mm})^3 = 0.0032$, $r = 0.147$ mm.

From Eq. (7.6), $f_n/f_{n0} = r/r_0$

$f_n = 91(0.147)/(0.5) = 29$ Hz.

7.6 (a) The circuit model for the system with pinch is given below.

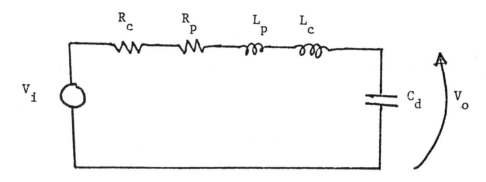

where

$$R_P = 8\eta L_{pinch}/\pi r_p^4$$
$$R_c = 8\eta (L - L_{pinch})/\pi r^4$$
$$L_p = \rho L_{pinch}/\pi r_p^2$$
$$L_c = \rho(L - L_{pinch})/\pi r^2$$
$$C_d = \Delta V/\Delta P$$

For a series R–L–C circuit

$$\zeta = R_{equivalent}/2 \sqrt{C_d/L_{equivalent}}$$

where

$$R_{equivalent} = R_c + R_p$$
$$L_{equivalent} = L_c + L_p$$

It is given that $\zeta = 0.7$ and $r_{p20°C} = 0.25R$. Assuming that the system is filled with H_2O @ 20°C

$$R_c = 8(0.001)(1 - L_{pinch})/\pi(0.046 \times 10^{-2})^4 = 5.69 \times 10^{10}(1 - L_{pinch})$$

$$R_p = 8(0.001) L_{pinch}/\pi(0.046 \times 10^{-2}/4)^4 = 1457 \times 10^{10} L_{pinch}$$

Then $R_{equivalent} = 5.69 \times 10^{10} + 1451 \times 10^{10} L_{pinch}$

and $L_c = (1 \times 10^3)(1 - L_{pinch})/\pi(0.046 \times 10^{-2})^2 = 1.5 \times 10^9(1 - L_{pinch})$

$$L_p = (1 \times 10^3)L_{pinch}/\pi(0.046 \times 10^{-2}/4)^2 = 24 \times 10^9 L_{pinch}$$

Blood Pressure and Sound

Then $L_{equivalent} = 1.5 \times 10^9 + 22.5 \times 10^9 L_{pinch}$

Thus we may solve for L_{pinch} from

$$\zeta = 0.7 = \frac{5.69 \times 10^{10} + 1451 \times 10^{10} L_{pinch}}{2} \sqrt{\frac{2.04 \times 10^{-15}}{1.5 \times 10^9 + 22.5 \times 10^9 L_{pinch}}}$$

Solution of the above equation gives

$$L_{pinch} = 0.14 \text{ m}$$

This may be checked by calculating ζ

$$R_p = 204 \times 10^{10}$$

$$R_c = 4.89 \times 10^{10}$$

Then

$$R_{equivalent} = 209 \times 10^{10}$$

and

$$L_p = 3.36 \times 10^9$$

$$L_c = 1.29 \times 10^9$$

Then

$$L_{equivalent} = 4.65 \times 10^9$$

$$C_d = 2.04 \times 10^{-15}$$

Now we can calculate ζ

$$\zeta = 209 \times 10^{10}/2 \sqrt{2.04 \times 10^{-15}/4.65 \times 10^9}$$

$$\zeta = 0.69$$

f_n can be determined

$$f_n = \frac{1}{2\pi} \sqrt{1/L_{equivalent} C_d}$$

$$f_n = \frac{1}{2\pi \sqrt{(4.65 \times 10^9)(2.04 \times 10^{-15})}} = 50 \text{ Hz}$$

or

$$\omega_n = 314 \text{ r/s}$$

(b) The frequency response for the system with and without the pinch is in (P. 7.4).

(c) The transient response for the system with and without the pinch when excited by a 100 mm Hg step input is shown below.

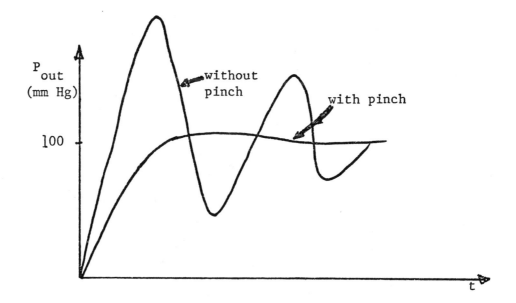

(d) Assuming that 10 harmonics gives a fairly good reproduction of a blood pressure waveform then the frequency response for a blood pressure sensor system for man should be essentially flat to 10 – 33 Hz, for dog 15 – 50 Hz and for shrew 120 – 220 Hz. From the frequency response curves we can see that the system should be adequate for recordings from man. Recordings from a dog with a high heart rate and recordings from the shrew will have attenuated high-frequency components.

7.7 We only need to know that a sound occurs and its amplitude. The 300-Hz wave is rectified and filtered by the envelope detector. This is modulated by an 80-Hz oscillator to yield waves that look like sounds and have the proper amplitude-time form.

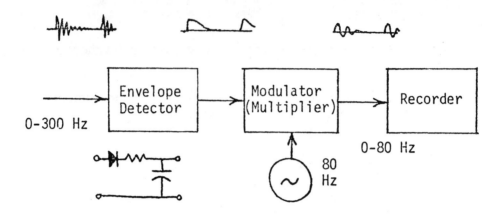

7.8

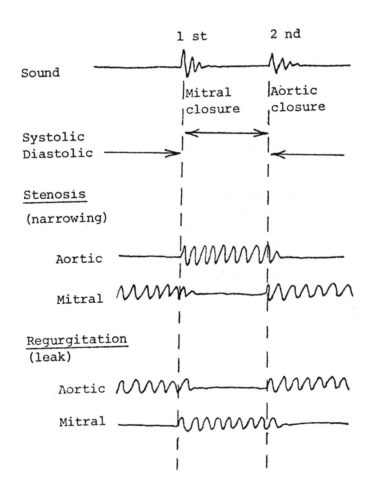

7.9 A block diagram for an automatic indirect blood pressure system is shown below.

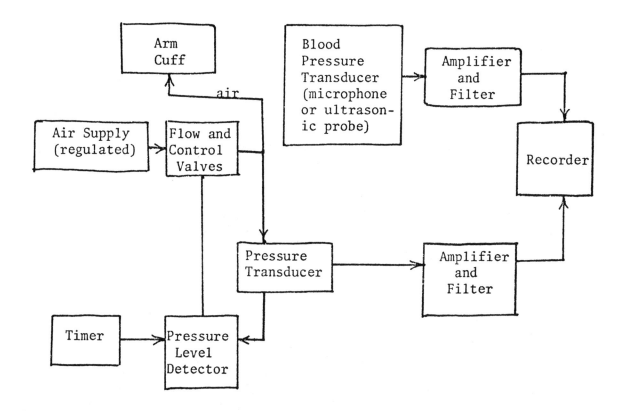

7.10 Modify Problem 7.9 for portable use. Replace the air supply with a battery-operated air pump. Add rechargeable batteries. Replace recorder with a battery-operated tape recorder or solid-state memory. Package to be carried in side pack with shoulder strap like a Holter monitor.

7.11 Due to the dehydration of the patient the veins will be only partially full and the jugular pulse will be essentially unobservable. As the veins become full with the infusion of physiological fluid the jugular pulse will reappear. The figure below shows the central venous pressure vs. time. The change in slope is a function of the mechanical properties of the vein

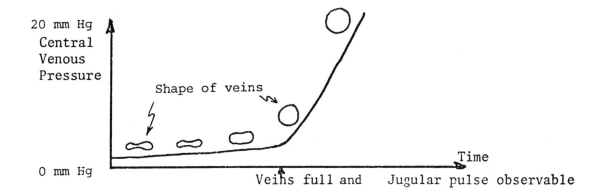

Blood Pressure and Sound

7.12 The kinetic energy K.E., term is $(1/2)\rho V^2$

$$\text{K.E.} = \frac{1}{2}(1050 \text{ kg/m}^3)(1.5 \text{ m/s})^2 = 1181 \text{ Pa}$$

or in terms of mm Hg

$$\text{K.E.} = 1181/133 = 9 \text{ mm Hg}$$

The orientation of the pressure catheter can significantly affect the measured aortic pressure. Under laminar flow conditions, this error may be decreased by moving the tip to wall where the average flow velocity is less.

7.13 Place piezoelectric pressure sensor on neck to measure carotid pulse. Place similar sensor on wrist to measure radial pulse. Use digital storage oscilloscope to display both waveforms and measure the difference in timing of pulse upstroke t for the two waveforms. Measure distance from aortic root to neck d_n. Measure distance from aortic root to wrist d_w.

$$\text{Velocity} = \text{distance/time} = (d_w - d_n)/t.$$

7.14 Under computer control, the air pump provides a ramp of pressure to the air bladder cuff that encircles the wrist. Below 80 mm Hg, the cuff pressure is insufficient to compress the artery. Above 80 mm Hg, the cuff pressure is higher than the blood pressure and relieves the hoop stresses of the wall. The sensor yields high peak-to-peak pulse pressures. Above 120 mm Hg, the artery is collapsed and yields no output. The computer stores all data, then searches for the maximal pulse pressure and reads the average sensor output at that time to indicate blood pressure.

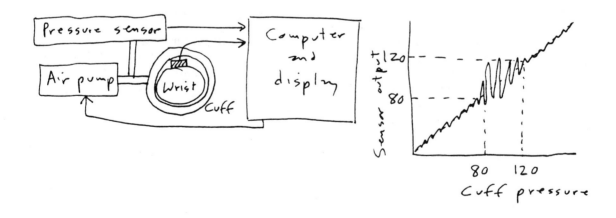

Chapter 8
Measurement of Flow and Volume of Blood

John G. Webster

8.1 From (8.1), $F = \dfrac{dm/dt}{\Delta C}$

F is the renal artery blood flow (clearance), ml/min.
ΔC is the PAH arterial concentration, mg/ml.
dm/dt is the PAH in the urine, mg/min.

clearance, ml/min = $\dfrac{\text{PAH in urine, mg/min}}{\text{PAH conc in art, mg/ml}}$

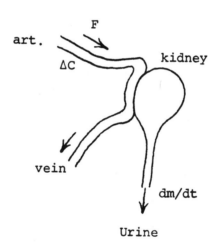

8.2 The definition of concentration is

$$C = \dfrac{m}{V} \text{ or } V = \dfrac{m}{C}$$

V is the circulating volume, ml
m is the amount injected, mg
C is the final concentration, mg/ml

8.3

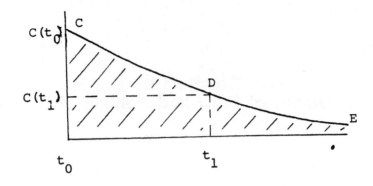

Assume $C(t) = C(t_0) \exp(-t/\tau)$

then $C(t_1) = C(t_0) \exp(-t_1/\tau)$

$$\frac{C(t_1)}{C(t_0)} = \exp(-t_1/\tau)$$

$$\ln \frac{C(t_1)}{C(t_0)} = -t_1/\tau$$

$$\tau = \frac{t_1}{\ln\left[\frac{C(t_1)}{C(t_0)}\right]}$$

Shaded area

$$A = \int_0^\infty C(t)dt = \int_0^\infty C(t_0) \exp(-t/\tau)dt$$

$$= C(t_0) \int_0^\infty C(t_0) \exp(-t/\tau)dt$$

$$= C(t_0)\tau = C(\tau_0)\frac{t_1}{\ln\left[\frac{C(t_1)}{C(t_0)}\right]}$$

8.4
$$F = \frac{Q}{\rho_b c_b \int_0^{t_1} \Delta T_b dt} = \frac{V_i T_i \rho_i c_i}{\rho_b c_b \int_0^{t_1} \Delta T_b dt}$$

$$= \frac{(10\text{ml})(-30\text{K})(1.005 \text{ g/ml})(4.172 \text{ J/(g·K)})}{(1.060 \text{ g/ml})(3.640 \text{ J/(g·K)})(-5.0 \text{ s·K})}$$

$$= 65.2 \text{ ml/s}$$

$(62.5 \text{ ml/s})(60 \text{ s/min}) = 3912 \text{ ml/min} = 3.9 \text{ l/min}$

8.5 Thermodilution. See equation (8.7).

8.6 Radiopaque. Optical absorption peak at 805 nm, the wavelength at which the optical absorption coefficient is independent of oxygenation.

8.7
$e = Blu$
$= (0.03 \text{ T})(0.015 \text{ m})(1 \text{ m/s})$
$= 0.00045 \text{ V} = 450 \text{ μV}$

8.8 Obtain steady flow in a flow rig:

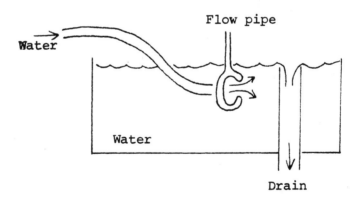

transiently short out the magnet current many times until a smooth transition is obtained (Calvert et al., 1975).

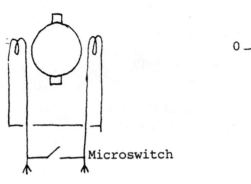
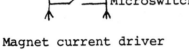
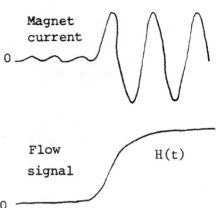

Differentiate the step response to obtain the impulse response h(t) = d/dt{H(T)}. Obtain the Fourier transform

$$H(\omega) = \int_{-\infty}^{+\infty} h(t) \exp(-j\omega t)\, dt$$

Obtain the real and imaginary parts

$$H(\omega) = R(\omega) + I(\omega)$$

Then the amplitude is

$$A(\omega) = [R(\omega)^2 + I(\omega)^2]^{1/2}$$

The phase is

$$\phi(\omega) = \tan^{-1}\left[\frac{I(\omega)}{R(\omega)}\right]$$

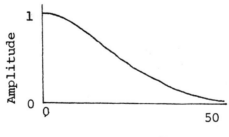

8.9

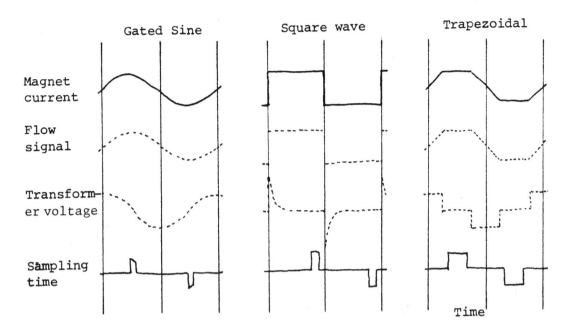

8.10 Figure 8.6 shows a quadrature suppression flowmeter. You can remove the quadrature suppression and it will still work because the in-phase demodulator rejects most of the quadrature (transformer) voltage that is 90° out of phase. From Fig. 3.17 we add the ring demodulator as the phase-sensitive (in-phase) demodulator.

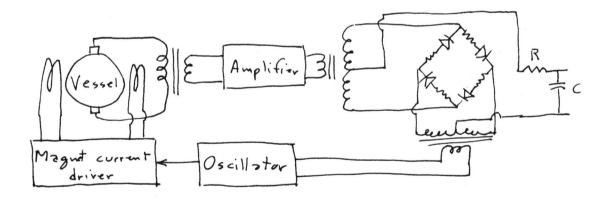

8.11

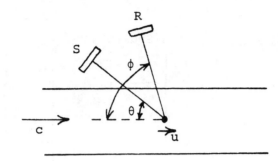

First, assume $\phi = \theta = 0°$. The equation for the Doppler principle is

$$f_c = f_s \left(\frac{c-u}{c} \right)$$

where

- f_s is frequency of source
- f_c is frequency received by cell
- c is velocity of sound
- u is velocity of cell

Now add the angle factor

$$f_c = f_s \left(\frac{c - u\cos\theta}{c} \right) \tag{8.11.1}$$

Note if $u\cos\theta = c$, $f_c = 0$, the frequency at receiver is

$$f_r = f_c \left(\frac{c}{c+u} \right)$$

Now add the angle factor

$$f_r = f_c \left(\frac{c}{c + u\cos\phi} \right) \tag{8.11.2}$$

Note if $u\cos\theta = c$, $f_r = f_c/2$. Now combine (8.11.1) and (8.11.2)

$$f_r = f_s \left(\frac{c - u\cos\theta}{c} \right) \left(\frac{v}{c + u\cos\phi} \right)$$

$$= f_s \left(\frac{c - u\cos\theta}{c + u\cos\phi}\right)$$

The Doppler frequency is

$$f_d = f_r - f_s = f_s \left(\frac{c - u\cos\theta}{c + u\cos\phi}\right) - f_s$$

$$= f_s \left(\frac{c - u\cos\theta}{c + u\cos\phi} - 1\right)$$

$$= f_s \left(\frac{c - u\cos\theta - c - u\cos\phi}{c + u\cos\phi}\right)$$

$$= f_s \left(\frac{-u(\cos\theta + \cos\phi)}{c}\right) \text{ for } c \gg u$$

If $\theta = \phi$, $f_d = -2f_s u\cos\theta/c$ \hfill (8.15)

8.12 Assume that the signal has been amplified so that its p-p value equals the ±10 V linear range for an op amp. Then our thresholds should be ±10/7 = ±1.4 V. Use the circuit shown in Fig. 3.6(a). Because the input is symmetric about zero, v_{ref} = 0 V.

Assume the op amp saturates at ±12 V. The p-p width of the hysteresis loop is four times the voltage across R_3 or 2.8/4 = 0.7. Assume R_3 = 1 kΩ. Then

$$\frac{R_2 + R_3}{R_3} = \frac{R_2 + 1\,k\Omega}{1\,k\Omega} = \frac{12}{0.7}$$

R_2 = 16.1 kΩ. Choose R_1 = 10 kΩ

8.13 Expand Fig. 8.11(b) as follows.

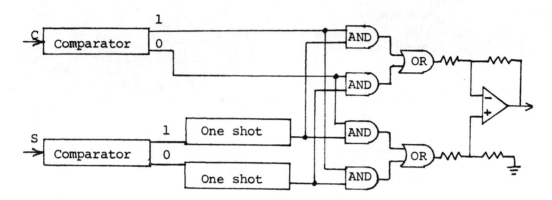

Expand Fig. 8.12(b) as follows.

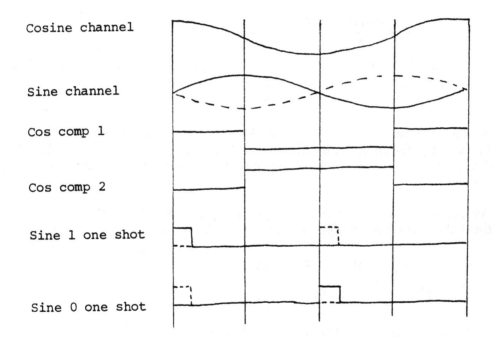

8.14

$$R_m = \frac{c}{2f_r} = \frac{1500 \text{ m/s}}{2(15000)\text{Hz}} = 0.05\text{m} = 50\text{mm}$$

$$u_m = \frac{c^2}{8f_0 \cos\theta \, R_m}$$

$$= \frac{(1500 \text{ m/s})^2}{8(8 \text{ MHz})(0.707)(0.050 \text{ m})} = 0.99 \text{ m/s}$$

8.15

+vel, R_3 warm, $v_r - 12$, relay pull.

−vel, R_4 warm, $v_r + 1$, no pull

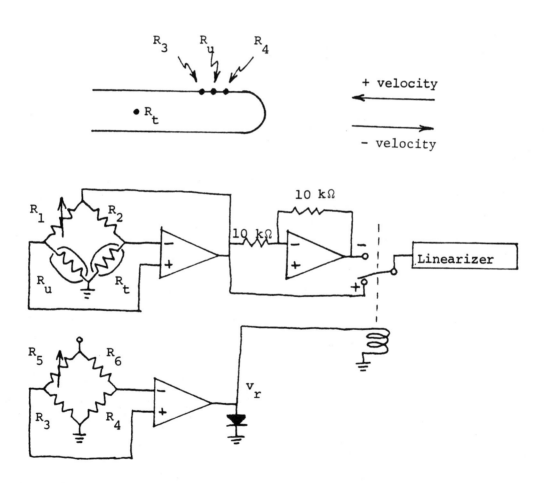

8.16 Temperature T increases because of compression.

$$P\{V\}^{1.4} = k$$
$$P = kV^{-1.4}$$
$$dP/dV = k(-1.4)V^{-2.4}$$
$$= -1.4\,kV^{-1.4}V^{-1}$$
$$= -1.4\,P(V^{-1})$$
$$= -1.4\,P/V$$
$$dV = -\frac{dP}{1.4}\frac{V}{P}$$
$$= \frac{(120)(200\text{ ml})}{1.4\,(101325)}$$
$$= -0.17\text{ ml}$$
$$dV\text{ (tissue)} = +0.17\text{ ml}$$

8.17 $F = dV/dT = 20\text{ ml}/9\text{ s} = 2.2\text{ ml/s}$.

8.18

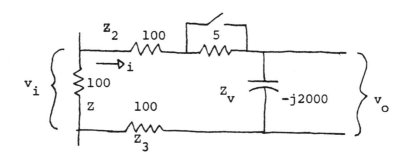

Assume v_i does not change when switch is opened.

Then $v_o \sim i \sim$ total impedance $Z_5 = Z_2 + Z_3 + Z_v$

$$Z_5 = 100 + 100 - j2000$$

$$|Z_6| = (200^2 + 2000^2)^{1/2} = 2010.0$$

After the switch is opened, the total impedance is $Z_7 = 100 + 5 + 100 - j2000$

$$|Z_7| = (205^2 + 2000^2)^{1/2} = 2010.5$$

$$\left|\frac{\Delta Z_6}{Z_6}\right| = \frac{0.5}{2010} \sim \frac{1}{4000} = 0.00025 = 0.025\%$$

This may be important because for many outputs

$$\left|\frac{\Delta Z}{Z}\right| \sim \frac{1}{1000} = 0.1\%$$

So the error would be $\dfrac{0.025\%}{0.1\%} = 0.25 = 25\%$

8.19

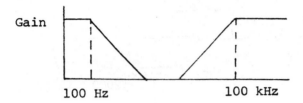

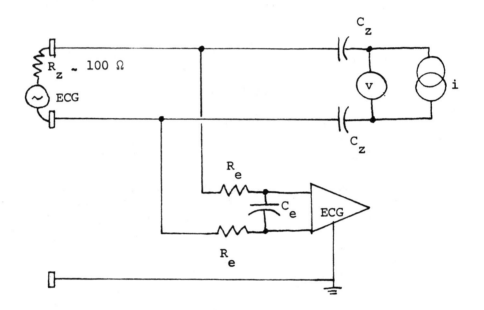

The 100 kHz impedance of the chest R_z is about 100 Ω. C_z should be as small as possible to block the ECG, but large enough to pass 100 kHz through the high-pass filter.

$$C_z = \frac{1}{2\pi fR} = \frac{1}{2\pi\, 10^4 (100)} = 0.16\ \mu F$$

Because two capacitors are in series and to allow a safety factor, choose $C_z = 1.0\ \mu F$. The phase sensitive demodulator would also block the ECG. Choose $2R_e = 10\ k\Omega$. Then to pass the 100 Hz of the ECG,

$$C_e = \frac{1}{2\pi fR} = \frac{1}{2\pi\, 100 (10^4)} = 0.16\ \mu F.$$

To allow a safety factor, choose $C_e = 0.1\ \mu F$. This low-pass filter will block the 100 kHz.

Chapter 9
Measurements of the Respiratory System

Frank P. Primiano, Jr.

9.1 • Models are the basis for our understanding of the physiological system, the measurement system and the interactions between them, i.e., models are formal statements of our ideas about the physiological and measurement systems.

  • They specify the variables related by the system under study.
  • They define the parameters of the system.
  • They are the basis for the design of experiments and measurements of the system.

9.2 From the total derivative of

$$P_A \left(\frac{V_L}{N_L}\right)^\alpha = K \tag{P.9.2.1}$$

we get

$$\frac{dP_A}{\alpha P_A} = -\frac{dV_L}{V_L} + \frac{dN_L}{N_L} \tag{P.9.2.2}$$

The lowest order approximation to this relation can be obtained by assuming that it will be evaluated about an operating point defined by $(\hat{P}_A, \hat{V}_L, \hat{N}_L)$. Then, for sufficiently small changes in any variable Y about this operating point,

$$dY \approx Y - \hat{Y} = y$$

Since $Y \approx \hat{Y}$ and $\alpha = 1$, (P 9.2.2) can be written as

$$\frac{p_A}{P_A} = -\frac{v_L}{V_L} + \frac{n_L}{N_L} \tag{P 9.2.3}$$

9.3 From P 9.2.2, we obtain

$$Cg_L = -\frac{\partial V_L}{\partial P_A} = \frac{V_L}{\alpha P_A} = \frac{2 \text{ liters}}{1(713 \text{ mm Hg})(1.36 \text{ cm H}_2\text{O/mm Hg})}$$

$$= .0021 \text{ liter/cm H}_2\text{O}$$

91

92 Measurements of the Respiratory System

9.4 Mechanical variables in (9.3a, b and c) can also be related by the barotropic relation given in (P 9.2.1) and by the dynamic mass balance of (9.2). The differential form of (P 9.2.1) written for the lungs is

$$\frac{p_ADRY}{P_ADRY} = -\frac{v_L}{V_L} + \frac{n_LDRY}{N_LDRY} \tag{P 9.4.1}$$

Integration of (9.2) written for dry gas assuming negligible net uptake over the time interval of interest and constant gas density yields

$$n_LDRY = \rho_{AWO}DRY \int q_{AWO}dt \tag{P 9.4.2}$$

Substituting

$$C_{gL} = \frac{V_L}{P_ADRY}, \quad \rho_LDRY = \frac{N_LDRY}{V_L} \approx \rho_{AWO}DRY$$

and (P 9.4.2) into (P 9.4.1) to eliminate N_LDRY and n_LDRY, produces

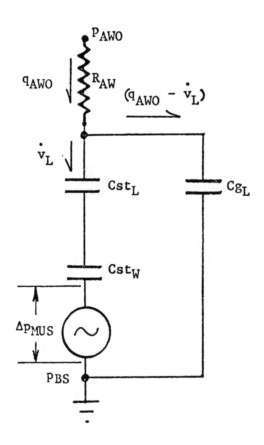

$$p_L = \frac{1}{C_{gL}}\left[-v_L + \int q_{AWO} dt\right] \qquad (P\ 9.4.3)$$

since $p_A DRY = p_A$ for a saturated mixture [see (9.29) and (9.30)]. Then (9.3a, b and c) and (P 9.4.3) can be represented by the equivalent circuit.

9.5 For a mixture of N_i mols of gas, i, in N mols of total gas, the mol fraction of i is

$$F_i = N_i/N$$

and the partial pressure of i is

$$P_i = F_i P = \frac{N_i}{N} P$$

so that

$$P_i V = N_i RT = \frac{n_i}{MW_i} RT$$

Where n_i is the mass of gas i, R is the universal gas constant, T is absolute temperature and V is volume.

Rearranging,

$$\frac{N_i}{V \cdot MW_i} = \frac{P_i}{RT} = \frac{N_i}{N} = \frac{P}{RT} = \frac{N_i/N}{V/N}$$

or

$$\left(\frac{n_i}{V}\right) \frac{(V/N)}{MW_i} = F_i$$

Let $\quad \gamma_i = (n_i/V) \propto mg/m^3$

$$(V/N) = 24.44\ dm^3/mol = 24.44 \times 10^{-3} m^3/mol$$

$$MW_i = MW_i(gm) = MW_i \times 10^3 mg$$

$$\therefore (\gamma_i\ mg/m^3) \times \frac{24.44 \times 10^{-3} m^3}{MW_i \times 10^3 mg} = F_i$$

$$\left(\gamma_i (mg/m^3) \times \frac{24.44}{MW_i}\right) \times 10^{-6} = F_i$$

or $\quad \gamma(mg/m^3) \times \frac{24.44}{MW} = F(ppm)$

94 Measurements of the Respiratory System

9.6 Molar density of X in a mixture is

$$\rho x = \frac{Nx}{V}$$

Molar fraction of X in a mixture is

$$Fx = \frac{Nx}{N}$$

By the ideal gas law

$$Px = \rho x RT = Fx \cdot P$$

$$\therefore \rho x = Fx \cdot \left(\mathcal{R}\frac{P}{T}\right)$$

If the output, e_o, of an instrument is a function of ρx, e.g.,

$$e_o = f_o(\rho x) = K_0 \rho x$$

then it will be the same function of Fx if the ratio P/T is constant, e.g.

$$e_o = f_o(K_1 Fx) = K_0 K_1 Fx = K_2 Fx$$

9.7

$$\frac{1}{Cst_{TR}} = \frac{1}{Cst_L} + \frac{1}{Cst_W} = \frac{1}{0.19} + \frac{1}{0.31}$$

$$\therefore Cst_{TR} = 0.1178 \text{ liter/cm } H_2O$$

9.8 1 liter/s through 100 tubes is 0.00001 m^3/s through one tube. In SI units, using (7.1) and (7.2),

$$\Delta P = RF = \frac{8\eta LF}{\pi r^4} = \frac{8(0.000018)(0.05)(0.00001)}{\pi (0.0005)^4}$$

$$= 367 \text{ Pa} = 3.74 \text{ cm } H_2O$$

9.9 The soda lime (NaOH) container acts as a flow resistor. During slow maneuvers or volume changes from one static state to another, pressure equilibrium throughout the spirometer and tubing system can be achieved. However, during fast maneuvers involving high flows, gas movement is impeded and pressure differences exist within the system. Consequently, the spirometer does not always indicate the gas volume at ambient pressure. If the canister is placed on the inlet side, expiratory flows would be attenuated but inspiratory flows might not be and vice-versa for the canister on the outlet side.

9.10 From the static mass balance of N_2, the following relation can be derived, assuming the experiment began at FRC:

$$V_L(t_1) = FRC = \frac{T_L}{T_S} \cdot \frac{F_S N_2(t_2) \, V_S(t_2)}{F_A N_2(t_1) - F_A N_2(t_2)(V_L(t_2)/FRC)} \quad (P\ 9.10.1)$$

Therefore, an estimate of FRC using 9.23 instead of (P 9.10.1) when the final volume is not FRC will lead to an answer which will have a fractional error of:

$$\therefore \frac{V_L - FRC}{FRC} = (1 - K) \cdot \frac{F_A N_2(t_2)}{F_A N_2(t_1) - F_A N_2(t_2)} \quad (P\ 9.10.2)$$

where $K = (V_L(t_2)/FRC)$

Note that for $F_A N_2(t_2) \ll F_A N_2(t_1)$, this error becomes negligible.

From the static He mass balance, assuming the lungs are at FRC at the end of the experiment:

$$V_L(t_2) = FRC = \frac{T_L}{F_S He(t_2)} \cdot \left[\frac{V_S(t_1) F_S He(t_1)}{T_S(t_1)} - \frac{V_S(t_2) F_S He(t_2)}{T_S(t_2)} \right] \quad (P\ 9.10.3)$$

Therefore an estimate of FRC using (9.24) instead of (P 9.10.3) when the final volume of the spirometer does not equal its original volume will lead to an answer which will have a fractional error of:

$$\frac{V_L - FRC}{FRC} = \frac{V_S(t_2) - V_S(t_1)}{\left[\dfrac{T_S(t_1) \cdot F_S He(t_1) \cdot V_S(t_1)}{T_S(t_2) \cdot F_S He(t_2)} \right] - V_S(t_2)} \quad (P\ 9.10.4)$$

9.11 The measurement ranges required from one PFT procedure to the next can differ by an order of magnitude or more. For example, flow during a single-breath N_2-washout is required to be much less than one 1 lps whereas peak flow during forced expirations can be in excess of 10 lps. In order to achieve acceptable resolution, signal to noise ratio and accuracy, it is sometimes necessary to use different instruments, with different operating ranges, to measure the same variable during different testing procedures. This is a particular consideration when evaluating commercial PFT systems which perform a variety of PFTs using the same set of measurements instruments.

9.12

$$V_S(t_2) = 5 \text{ liter} + 7 \text{ liter}$$

$$F_S N_2(t_2) = .026$$

$$F_A N_2(t_1) - F_A N_2(t_2) = 0.1$$

$$T_S = 303 \text{ K}$$

$$T_L = 310 \text{ K}$$

$$V_L = \frac{T_L}{T_S} \cdot \frac{F_S N_2(t_2) V_S(t_2)}{F_A N_2(t_1) - F_A N_2(t_2)}$$

$$V_L = \frac{310}{303} \cdot \frac{.026 \times 12}{0.1} = 3.19 \text{ liter}$$

9.13 The amount of N_2 in the lungs at RV before a VC inspiration of pure O_2 to TLC is $F_A N_2(t_0) \cdot RV$ where $RV = (TLC - VCI)$. This amount remains the same as a result of the O_2 inspiration since it is assumed that negligible N_2 is added to, or taken from, the lungs during this inspiration. Therefore,

$$F_A N_2(t_0) \cdot (TLC - VCI) = F_A N_2(t_1) \cdot (TLC - V_D) \quad \text{(P 9.13.1)}$$

where $TLC - V_D$ is the alveolar volume and t_1 is the time at which the lungs reached TLC. Note that (P 9.13.1) is not a valid relation in the presence of significant ventilation inhomogeneity. The alveolar molar fraction of N_2 at t_1 can be estimated from the succeeding expiration since the cumulative expired volume of N_2 will be that resulting from the change in alveolar volume during the expiration, i.e., $VCE - V_D$. Therefore,

$$F_A N_2(t_1) \cdot (VCE - V_D) = \int_{t_1}^{t_2} F_E N_2 \, Q_{AWO} dt \quad \text{(P 9.13.2)}$$

These equations, (P 9.13.1) and (P 9.13.2), can be rearranged to yield those given in problem 9.13. V_D and $F_A N_2(t_0)$ must be independently estimated to solve these equations. Conversely, if $F_A N_2(t_1)$ is independently estimated, then (P 9.13.2) is equivalent to (9.47) and can be used to estimate V_D. Therefore, (9.47) cannot be used to estimate V_D for use in (P. 9.13.2) since they become redundant, i.e., $F_A N_2(t_1) \equiv \widehat{F}_E N_2(III)$.

9.14
$$V_L = \frac{V_S(t_1)}{F_SHe(t_2)} \left[\frac{T_L}{T_S(t_1)} F_SHe(t_1) - \frac{T_L}{T_S(t_2)} F_SHe(t_2) \right]$$

$$T_L = 37 + 273 = 310°K$$

$$T_S(t_1) = 25 + 273 = 298°K$$

$$T_S(t_2) = 305°K$$

$$V_L = \frac{10}{.04} \left[\frac{310}{298} \times .05 - \frac{310}{305} \times .04 \right]$$

$$= 2.84 \text{ liters} \tag{P 9.14.1}$$

Note that this estimate of FRC is much more dependent on the temperature factors than the estimate from an N_2 washout.

9.15 From (9.31) and (9.32) in Section 9.5:

$$V_{TG} = -(P_{atm} - P_AH_2O) \frac{V_B}{(\alpha_B)P_B} \frac{(dP_B)_0}{(dP_M)_0}$$

$$= -(760 - 47) \frac{(1000 - 100)(-.1)}{1.4 \cdot 760 \cdot 30}$$

$$= 2.01 \text{ liters} \tag{P 9.15.1}$$

9.16 Following the derivation of (9.30) in section 9.5, the mass balance on the lungs given after (9.29a) can be rewritten to include net gas exchange with the blood:

$$dN_LDRY = \rho_{AWO}DRY \cdot Q_{AWO}dt - \dot{U}_bDRYdt \tag{P 9.16.1}$$

Substituting this into (9.29a) along with the mass balance on the box, and considering the points made in the paragraph following (9.29b) yields

$$\frac{V_L}{P_ADRY} dP_A = -\left[\frac{V_B}{\alpha_B P_B} dP_B + \left(\frac{T_B}{T_{AWO}} - \frac{P_{AWO}DRY \cdot T_L}{P_ADRY \cdot T_{AWO}} \right) Q_{AWO}dt + \frac{\dot{U}_bDRYdt}{\rho_LDRY} \right]$$

$$\tag{P 9.16.2}$$

9.17 Following the derivation of (9.30) in section 9.5, the constraint given by (9.27) can be rewritten

$$dV_L = dV_1 + dV_2 = -dV_B$$

Then (9.29a) can be rewritten

$$\frac{-V_1}{P_1 DRY} dP_1 DRY - \frac{V_2}{P_2 DRY} dP_2 DRY + \frac{dN_1 DRY}{\rho_1 DRY} + \frac{dN_2 DRY}{\rho_2 DRY}$$

$$= \frac{V_B}{\alpha_B P_B} dP_B - \frac{dN_B}{\rho_B}$$

and mass balances on the lungs and box yield

$$dN_1 DRY + dN_2 DRY = \rho_{AWO} DRY \cdot Q_{AWO}\, dt$$

$$dN_B = -\rho_{AWO} Q_{AWO}\, dt$$

If $\rho_1 DRY \approx \rho_2 DRY \approx \rho_L DRY$ and $P_1 DRY \approx P_2 DRY \approx P_A DRY$, then

$$-\frac{V_1 dP_1 DRY}{P_A DRY} - \frac{V_2 dP_2 DRY}{P_A DRY} = \frac{V_B}{\alpha_B P_B} dP_B + \frac{\rho_{AWO}}{\rho_B} Q_{AWO}\, dt$$

$$-\frac{\rho_{AWO} DRY}{\rho_L DRY} Q_{AWO}\, dt$$

or

$$\frac{V_L}{P_A DRY} \sum_{i=1}^{2} \frac{V_i dP_i DRY}{V_L} =$$

$$-\left[\frac{V_B}{\alpha_B P_B} dP_B + \left(\frac{\rho_{AWO}}{\rho_B} - \frac{\rho_{AWO} DRY}{\rho_L DRY}\right) Q_{AWO}\, dt\right] \quad (P\ 9.17.1)$$

Define $\sum_i \frac{V_i dP_i DRY}{V_L} \equiv d\widehat{P}_A$, a representative weighted, mean alveolar pressure change. Then (P 9.17.1) reduces to the same form as (9.29b). It can be seen that changes in box pressure are related to a weighted average of all pressure changes in the lung, not to a single, unique "alveolar" pressure change.

9.18 (a) Place the subject in a TBP such as that in Figure 9.8. With the subject's mouth and nose blocked, oscillate the calibration pump and measure the volume displacement of the pump, V_P, and the amplitude of the resulting box pressure change, P_B. The ratio of these values is the acoustic compliance of the gas in the box (see p. 490):

$$V_P/P_B = V_B/\alpha_B P_B$$

For the TBP being used, α_B can be evaluated with this same procedure but with objects of known volume in place of the subject in the plethysmograph. (α_B should approach 1.4 for a well sealed and insulated box). Then the volume of the gas space around the subject in the TBP, V_B, can be calculated and subtracted from the known empty volume of the TBP, V_P, to yield the total volume of the subject's body (see 9.26):

$$V_{body} = V_{TIS} + V_L = V_P - V_B = V_P - \alpha_B P_B[V_p/P_B]$$

(b) With the subject still in the TBP, follow the procedure described in the section entitled "Volume of Gas Within the Thoracic Cavity by Total-Body Plethysmograph," p. 489–491, and use 9.32 to evaluate V_{TG} as a measure of V_L.

(c) Weigh the subject (unclothed). Calculate the subject's tissue density as:

Tissue Density = Body Mass/V_{TIS}
= (Body Weight/g)/($B_{body} - V_L$)

where g is the acceleration of gravity.

9.19• Magnetometers: chest wall diameter changes.
- Strain gages: chest wall circumference changes.
- Inductive plethysmograph: cross-sectional area changes.
- Total Body Plethysmograph: thoracic volume changes.

Note: Electrical impedance "plethysmograph" instruments utilizing surface electrodes only yield measures of breathing activity and are more reliable as breathing rate monitors than as indicators of volume change.

9.20 Three parameters discussed in this chapter yield measures of large airway obstruction. They are: (1) airway resistance, R_{AW}, obtained from plethysmographic measurements; (2) peak expiratory flow, PEF, from a forced expiratory effort beginning at total lung capacity and (3) the maximal voluntary ventilation, MVV. These are discussed in Section 9.6 in the paragraphs entitled *Dynamic Mechanics*. Another related parameter is specific airway resistance, SR_{AW}, discussed in the same section in the paragraphs entitled *Large Amplitude Volume and Flow Behavior of the Ventilatory System*.

9.21 The three tests for "small" airways disease mentioned in this chapter are: (1) a decreasing pulmonary dynamic compliance, $Cdyn_L$, with increasing breathing frequency coexistent with a normal R_{AW} and normal Cst_L; (2) parameters associated with the latter part of the forced vital capacity maneuver, e.g., FEF50%/FVC, FEF 25–75%, curvature of the MEFV curve or timed spirogram (See Section 9.6, paragraphs entitled *Forced Expiratory Flows and Forced Expiratory Volume* and *The Patterns of Obstruction and Restriction*); (3) the closing volume, CV, and closing capacity, CC (see Section 9.8, paragraph entitled *Gas-Phase Transport*).

9.22 Insert a balloon-tipped catheter into the thoracic esophagus of the anesthetized animal through the mouth and connect it to one side of a differential pressure gage. After the animal is paralyzed its respiration should be maintained by a ventilator connected to its trachea through an endotracheal tube inserted through the mouth. The second side of the differential pressure gage should be connected to a tube passed through the endotracheal tube and exposed to hydrostatic pressure in the trachea below the glottis. The volume change of the lungs can be estimated either by integrating flow through the endotracheal tube or by placing the animal in a flow-displacement plethysmograph with the endotracheal tube communicating to the outside of the box.

By regulating the gas put into or withdrawn from the lungs by the ventilator, the ventilatory system of the animal can be made to undergo stepwise static changes in volume and a statically-determined ΔP–V curve can be defined for the chestwall and lungs; chestwall and lung static compliances (9.7) can then be evaluated. If the ventilator can produce (quasi) sinusoidal gas flows, an effective time constant, τ, can be computed as

$$\tau = \frac{\Delta p_{TP} Cst_L - v_L}{\dot{v}_L}$$

where the resting values of ΔP_{TP}, V_L, and $\dot{V}_L$ are taken for the same instant in time.

It should be noted that the time constant of the nearly exponential decay of the volume of the system after release from a given volume above FRC down to FRC during paralysis will not be the time constant of the lungs alone but that of the total ventilatory system.

9.23

From (9.11)

$$R_{AW} = \frac{(P_{AWO} - P_{PL}) - \frac{1}{C_{STL}} v_L}{\dot{v}_L}$$

$$= \frac{(P_{AWO} - P_{PL})(t_1) - (P_{AWO} - P_{PL})(t_2) - \frac{1}{C_{STL}}(v_L(t_1) - v_L(t_0))}{\dot{v}_L(t_1) - \dot{v}_L(t_0)}$$

Measure volume and flow of gas inspired and expired at the airway opening (as an estimate of v_L and $\dot{v}_L$, respectively) and $(P_{AWO} - P_{esoph})$ (as an estimate of $(P_{AWO} - P_{PL})$) simultaneously and continuously. Identify t_1 and t_2 as two instants at which $v(t_1) = v(t_0)$ and $\dot{v}(t_1) \neq \dot{v}(t_0)$.

Then:
$$R_{AW} = \frac{(P_{AWO} - P_{PL})(t_1) - (P_{AWO} - P_{PL})(t_0)}{\dot{v}_L(t_1) - \dot{v}_L(t_0)}$$

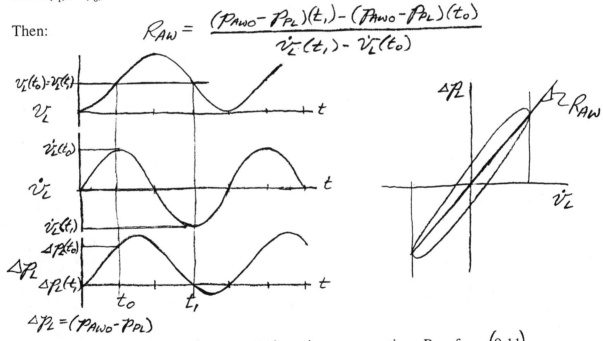

$\Delta P_L = (P_{AWO} - P_{PL})$

R_{AW} from the plethysmograph measures just airway properties. R_{AW} from (9.11) would include any flow resistive properties of the lung tissue. These properties were assumed to be negligibly small in the derivation of (9.11).

9.24

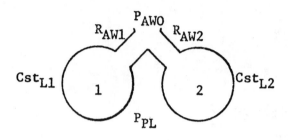

Assume $\dot{v}_L = q_L$

$$\Delta p_L = p_{AWO} - p_{PL} = \frac{1}{Cst_{L1}} v_1 + R_{AW1}\dot{v}_1 \quad (P\ 9.24.1)$$

$$\Delta p_L = \frac{1}{Cst_{L2}} v_2 + R_{AW2}\dot{v}_2 \quad (P\ 9.24.2)$$

$$v_L = v_1 + v_2$$

$$\dot{v}_L = \dot{v}_1 + \dot{v}_2 \quad (P\ 9.24.3)$$

Then

$$\Delta p_L + K_1 \Delta \dot{p}_L = \gamma_0 v_L + \gamma_1 \dot{v}_L + \gamma_2 \ddot{v}_L \quad (P\ 9.24.4)$$

where

$$K_1 = \frac{(R_{AW1} + R_{AW2})\, Cst_{L1} Cst_{L2}}{Cst_{L1} + Cst_{L2}}$$

$$\gamma_0 = \frac{1}{Cst_{L1} + Cst_{L2}} = \frac{1}{Cst_L}; \quad \gamma_1 = \frac{R_{AW2} Cst_{L2} + R_{AW1} Cst_{L1}}{Cst_{L1} + Cst_{L2}}$$

$$\gamma_2 = \frac{R_{AW1} R_{AW2} Cst_{L1} Cst_{L2}}{Cst_{L1} + Cst_{L2}}$$

After Laplace transformation (P 9.24.4) becomes

$$\frac{\Delta p(s)}{\dot{v}(s)} = \frac{\gamma_0 + \gamma_1 s + \gamma_2 s^2}{s(1 + K_1 s)} \quad (P\ 9.24.5)$$

substituting $s = j\omega$ into (P 9.24.5) for sinusoidal forcing, yields $Z_L = \frac{\Delta p(\omega)}{\dot{v}(\omega)}$. Rearrangement of Z_L into real and imaginary parts yields

$$Ceff_L = \frac{\omega^2(\tau_2 Cst_{L1} + \tau_1 Cst_{L2})^2 + (Cst_{L1} + Cst_{L2})^2}{\omega^2(\tau_1^2 Cst_{L2} + \tau_2^2 Cst_{L1}) + (Cst_{L1} + Cst_{L2})} = \frac{-1}{\omega \mathrm{Im} Z_L} \quad (P\ 9.24.6)$$

Measurements of the Respiratory System

and

$$\text{Reff}_L = \frac{\omega^2 \tau_1 \tau_2 (\tau_2 \text{Cst}_{L1} + \tau_1 \text{Cst}_{L2}) + (\tau_1 \text{Cst}_{L1} + \tau_2 \text{Cst}_{L2})}{\omega^2 (\tau_2 \text{Cst}_{L1} + \tau_1 \text{Cst}_{L2})^2 + (\text{Cst}_{L1} + \text{Cst}_{L2})^2} \quad (P\ 9.24.7)$$

where $\tau_i + R_{AWi} \text{Cst}_{Li}$

9.25 Those whose molecules possess an electric dipole moment, i.e., those made up of dissimilar atoms, e.g., CO_2, CO, N_2O, H_2O, etc. IR radiant energy is not absorbed by symmetrical molecules such as O_2, N_2, H_2, He, etc.

9.26 P_t = power transmitted; P_0 = power entering; P_a = power absorbed.

$$P_a = P_0 - P_t = P_0(1 - e^{-aLC}) \approx P_0 aLC \text{ for low absorption } P_a/P_0 < 0.1$$

9.27
$$Q = \hat{u} \times \frac{\pi D^2}{4}$$

where $\hat{u}$ = mean velocity of the gas flowing through a tube having a diameter D.

Also,
$$\hat{u} = \frac{L}{t}$$

where L = length of the tube. This is traversed in time t, the transit delay time.

$$\therefore Q = \frac{L}{t} \frac{\pi D^2}{4}$$

$$L = \frac{4Qt}{\pi D^2} = \frac{4 \times 10 \text{cm}^3 \text{min} \times 0.25}{3.14 \times 60 \text{s/min} \times (D(\text{cm}))^2} = \frac{0.0424}{D^2} \text{ (cm)}$$

9.28 • Mass spectrometer: mass/charge ratio.
- Raman spectrometer: wavelength shift of scattered light.
- Paramagnetic balance-type sensor: paramagnetism of oxygen.
- Paramagnetic differential pressure-type sensor: paramagnetism of oxygen.
- Magnetoacoustic sensor: paramagnetism of oxygen.

9.29 See Figure 9.17b and Section 9.8, paragraph entitled *Gas-Phase Transport*.

9.30 Assume that at the end of a vital capacity, VC, inspiration of pure O_2, the anatomical dead space volume, $anatV_D$, is filled with pure O_2 and all the nitrogen in the lungs is in the alveolar volume at a representative alveolar molar fraction $\widehat{F}_EN_2(III)$ obtainable from phase III of the washout curve. Also assume that during an expiration, $anatV_D$ does not change so that the alveolar volume undergoes a change in volume equal to VC. Then, all the N_2 expired during a VC expiration must come from the alveoli and will either be measured in the expirate, $\int F_EN_2 dv_S$, or remain in the dead space $anatV_D \cdot \widehat{F}_EN_2(III)$. Therefore,

$$VC \cdot \widehat{F}_EN_2(III) = \int F_EN_2 dv_S + anatV_D \cdot \widehat{F}_EN_2(III) \qquad (P\ 9.30.1)$$

This can be arranged to yield (9.44).

To evaluate effective or physiologic dead space, V_D, assume that all of the CO_2 leaving the lung, computed as the product of the mixed expired molar fraction and the tidal volume, $\widehat{F}_ECO_2 \cdot V_T$ must have come from the change in alveolar volume during the breath $(V_T - V_D)$. The gas from the alveolar volume, prior to the tidal volume expiration, had a molar fraction of $\widehat{F}_ACO_2$. A mass balance yields

$$\widehat{F}_ECO_2 \cdot V_T = \widehat{F}_ACO_2(V_T - V_D) \qquad (P\ 9.30.2)$$

Multiplying both sides by P_{atm}, we obtain

$$\widehat{P}_ECO_2 \cdot V_T = \widehat{P}_ACO_2(V_T - V_D) \qquad (P\ 9.30.3)$$

Mixed alveolar CO_2 partial pressure, $\widehat{P}_ACO_2$, can be estimated from systemic arterial CO_2 partial pressure, $\widehat{P}_aCO_2$.

Anatomical V_D is an estimate of a dead space defined only for gas-phase transport and reflects distribution by convection and some gas-phase diffusion. Effective or physiologic V_D is affected not only by gas-phase transport but also by diffusion processes between the alveoli and blood. The values found by each are comparable in normal lungs when ventilation and perfusion are similarly distributed. Otherwise, $anatV_D$ will be lower than effective V_D.

9.31 From the definition of diffusing capacity for a gas X, D_LX, the rate of uptake of X by the blood (expressed as an equivalent volume uptake, U_bX/ρ_ADRY) is proportional to the partial pressure difference of X between the alveoli and the pulmonary capillary blood:

$$\frac{\dot{U}_bX}{\rho_ADRY} = D_LX(P_AX - P_bX) \qquad (P\ 9.31.1)$$

The gas routinely used is CO and P_bCO can be assumed to be zero. For a breathholding experiment, it can be assumed that the alveolar space is uniformly filled at t_1 with a mixture containing CO with a molar fraction of $F_ACO(t_1)$ on a dry gas basis and the alveolar volume, V_A, and total dry gas density, ρ_ADRY, do not change from t_1 to t_2. Then for a mass balance for CO in the alveolar space (see 9.2):

$$\frac{d(N_ACO)}{dt} = -\dot{U}_bCO$$

$$\frac{d(\rho_ADRY \cdot F_ACO \cdot V_A)}{dt} = -D_LCO \cdot F_ACO(P_{atm} - P_AH_2O)\rho_ADRY$$

Since ρ_ADRY and V_A are essentially constant during the breathhold:

$$\dot{F}_ACO \cdot V_A = -D_LCO \cdot F_ACO \cdot (P_{atm} - P_AH_2O) \qquad (P\ 9.31.2)$$

This is a homogeneous, linear, first-order differential equation which can be solved by integration for F_ACO:

$$\ln\frac{F_ACO(t_2)}{F_ACO(t_1)} = \frac{-D_LCO \cdot (P_{atm} - P_AH_2O)(t_2-t_1)}{VA} \qquad (P\ 9.31.3)$$

This can be rearranged to yield (9.44) or can be solved for D_LCO explicitly:

$$D_LCO = \frac{VA}{(t_2-t_1)(P_{atm} - P_AH_2O)} \ln\left(\frac{F_ACO(t_2)}{F_ACO(t_1)}\right)^{-1} \qquad (P\ 9.31.4)$$

A discussion of the manner in which the experiment can be performed and how VA and $F_ACO(t_1)$ and $F_ACO(t_2)$ are obtained is given in the paragraphs before and after (9.46).

Chapter 10
Chemical Biosensors

Robert A. Peura

10.1 Section 10.2 states that internal impedance of the pH electrode is in the 10- to 100-MΩ range. Thus we need an amplifier with extremely high input impedance and extremely small bias current. This suggests an FET op amp which has specifications for extremely low bias current and extremely low offset voltage drift. This can be connected as a noninverting amplifier with a gain of 100.

10.2 See Fig. 10.3. CO_2 diffuses through a semipermeable membrane, which blocks ions. It dissolves in water to form carbonic acid, which dissociates to form H^+ ions. These are measured by a conventional pH electrode.

10.3 As in the discussion for the oxygen electrode, the material and thickness of the semipermeable membrane determine the time it takes for the CO_2 to diffuse through it and hence the response time.

10.4 Place a cup-shaped contact lens on the eye, which is filled with a known concentration of oxygen (volume of O_2/volume in contact lens) in physiologic saline. The inner surface of the contact lens which is in contact with the eye should be permeable to O_2. The O_2 flux into the eye is determined by the following.

$$\text{Flux} = \frac{Q}{A \cdot t}$$

where

Q = volume of O_2

A = contact area

t = time

$$Q = V \Delta C$$

where

V = volume in contact lens

ΔC = difference in concentration of O_2 between initial value and value after one hour.

Thus measure O_2 concentration in solution initally and after one hour using P_{O_2} electrode. Note that the P_{O_2} electrode measures the partial pressure of O_2, which is directly proportional to the O_2 concentration in physiologic saline.

10.5 From a −15-V power supply use a 14300-Ω and 700-Ω resistor voltage divider to yield −0.7 V to bias the Pt electrode. Feed the Ag/AgCl electrode output into an FET current-to-voltage converter with a feedback resistor = V/I = 10 V/250 nA = 40 MΩ.

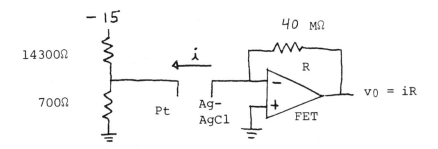

10.6 The material and thickness of the semipermeable membrane determine the time it takes for the O_2 to diffuse through it and hence the response time.

10.7 Use two oxygen sensors with equations given in (10.10) and (10.11). Immobilize glucose oxidase in front of only sensor 1. Then

$$\text{Glucose} + O_2 \Rightarrow \text{Gluconic acid} + H_2O_2 \qquad (10.18)$$

Sensor 2 senses only oxygen, as described in the text. Glucose is determined as a function of the difference between the two measurements.

10.8 The source illumination is split into two parts. One part passes through the sample and is attenuated. The other reference part passes directly to the detector. The instrument takes the ratio of sample to reference to make the measurement. The advantage is that the source can vary without changing the ratio. Fig. 10.24 shows an example of an intravascular optical fluorescence blood gas analyzer that uses a double beam.

10.9 $\quad\text{pH} = -\log_{10}[H^+]$
$\qquad\quad = -\log_{10}[40 \times 10^{-9}]$ moles/liter
$\qquad\quad = -[1.6 - 9.0] = 7.4$

So pH is in the normal range 7.38 – 7.44. But the PCO_2 is 60 mm Hg, which is high compared to the normal value of 40 mm Hg. Hence the patient has decreased overall ventilation. If the patient increased ventilation to lower PCO_2 to normal, this would remove CO_2 and carbonic acid and increase pH. Thus the patient also has metabolic alkalosis. But the metabolic alkalosis has been compensated for by reduced ventilation.

Chapter 11
Clinical Laboratory Instrumentation

Lawrence A. Wheeler

11.1 Photometers and colorimeters use filters that have fixed wavelengths. They are simpler, economical, and less accurage. A monochromator uses a slit plus a prism or a diffraction grating. It can sweep through a band of wavelengths and thus is more versatile. It is more complex, expensive, and more accurate.

11.2 From (11.3)

$$A_s = 2 - \log \% T$$
$$= 2 - \log 20$$
$$= 2 - 1.30 = 0.70$$
$$A_u = 2 - \log 30$$
$$= 2 - 1.48 = 0.52$$

From (11.5)

$$C_u = C_s \left(\frac{A_s}{A_u}\right) = 8 \frac{0.52}{0.70} = 5.9 \text{ g/dl}$$

11.3

$$A_1 = 2 - \log 79.4 = 2 - 1.90 = 0.10$$

$$A_2 = 2 - \log 39.8 = 2 - 1.60 = 0.40$$

$$A_3 = 2 - \log 31.6 = 2 - 1.50 = 0.50$$

$$A_4 = 2 - \log 20 = 2 - 1.30 = 0.70$$

Yes, see straight line in curve

$$A_5 = 2 - \log 35 = 2 - 1.544 = 0.456$$

$$C = 9 \text{ mg/dl (from curve)}$$

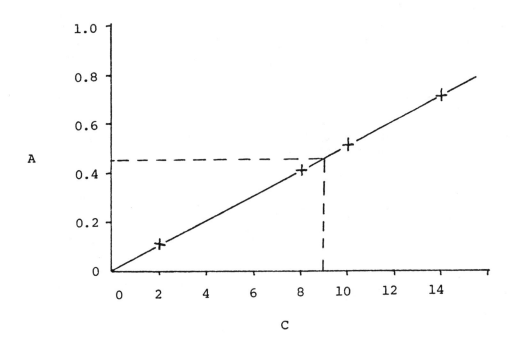

110 Clinical Laboratory Instrumentation

11.4

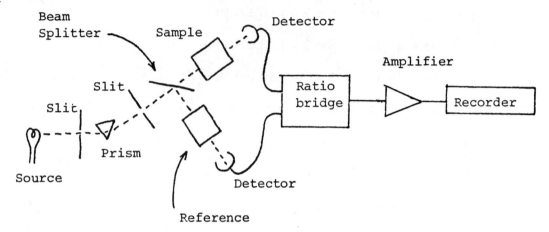

The prism disperses the wavelengths, the slit picks one single wavelength at a time. The sample is compared against a known reference and the ratio of their transmittances recorded as a function of wavelength. Variations caused by the source, slit width, etc. are common to both channels and do not cause error in the ratio.

11.5 Direct measurement of fluorescence. Small number of substances fluoresce. The emission spectra are different for each substance.

11.6
1. What is the ordering frequency of the various tests in the chemistry area?
2. Which tests are commonly ordered together?
3. Can the instrument efficiently perform appropriate panels of the high-volume tests?
4. What are reagent costs?
5. How fast must tests be done?
6. How much flexibility is required?
7. Will a computer be used?

11.7 $\text{RBC count} = \dfrac{10 \text{ HCT}}{\text{MCV}} = \dfrac{10(40)}{90} = 4.44 \text{ million/}\mu l$

$\text{Hb} = \dfrac{\text{MCH(RBC count)}}{10} = \dfrac{30(4.44)}{10} = 13.3 \text{ g/dl}$

$\text{MCHC} = 100 \dfrac{\text{Hb}}{\text{HCT}} = 100 \dfrac{13.3}{40} = 33.3\%$

11.8 Bias resistors provide constant current. Input capacitors block the large dc. Amplifiers are followed by a threshold detector (Th) that puts out an equal-area pulse for each detected cell. These are integrated to form voltages 1, 2, 3. In voting logic circuit resistors are 10 kΩ and take averages. The difference amplifier takes the difference between a voltage and the average of the other two. If its magnitude exceeds a threshold, it opens its switch and does not contribute to the output. If all three magnitudes are high, AND gate indicates an error.

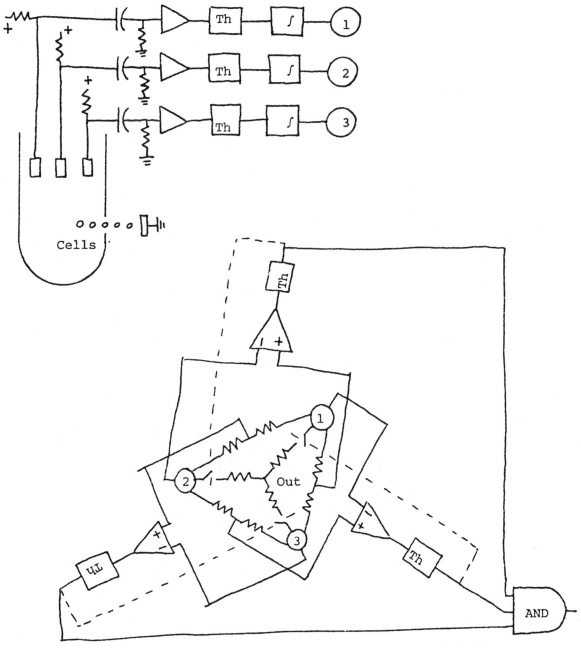

Chapter 12
Medical Imaging Systems

Melvin P. Siedband

12.1 From example 12.1 the standard TV bandwidth is 4.63 MHz for a frame time of 1/30 s. If we are allowed only 0.003 MHz, the transmission time will be

$$t = \left(\frac{4.63}{0.003}\right)\left(\frac{1}{30}\right) = 51 \text{ s}.$$

12.2 Standard system: Df = (480)(640)(67)/[(2)(0.9)(0.9)] = 12.7 MHz
High resolution: Df = (1024)(1280)(67)/[(2)(0.9)] = 48.8 MHz
Interlaced scanning would halve the bandwidth

12.3 The worst orientation is shown. There are 8 vertical objects (black squares). As in Fig. 12.1, 2n = 16 scanning lines (shown dashed) are not enough. Therefore use $2n\sqrt{2} = 22.6 \simeq 23$ lines. There are 8 horizontal objects to use 8 cycles. Total cycles = 23 × 8 = 184 cycles/image.

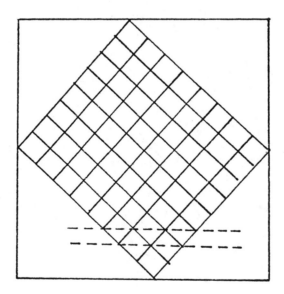

12.4 Use (12.2)

$p(K;m) = e^{-m}m^K/K!$
$p(0;3) = e^{-3}3^0/0! = 0.043$
$p(1;3) = e^{-3}3^1/1! = 0.13$
$p(2;3) = e^{-3}3^2/2! = 0.19$
$p(3;3) = e^{-3}3^3/3! = 0.19$
$p(4;3) = e^{-3}3^4/4! = 0.145$
$p(5;3) = e^{-3}3^5/5! = 0.081$

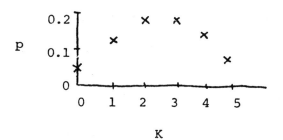

12.5 $\bar{M} = 17/25 = 0.68$

$J = \bar{M}(N)^{1/2} = 0.68(25)^{1/2} = 3.4$

From a table of single-ended probabilities of the normal (Gaussian) distribution for $J = 3.4$, $p = 0.0003$.

Exceedances $= 0.0003(10^4) = 3$

12.6 Solve as (12.9),

$C = 1/\text{gray levels} = 1/6.$
$N = A[7.2/d(C - 0.05)]^2$
$= 200^2[7.2/(1)(0.167 - 0.05)]^2$
$= 15{,}150{,}000$

12.7 From (12.7)

$$N_e = \int_0^\infty \left(\frac{1}{1+2f}\right)^2 df$$

$$\frac{1}{2}\left(\frac{1}{1+2f}\right)\bigg|_0^\infty = 0 - \left(-\frac{1}{2}\right) = \frac{1}{2} \text{ cy/mm}$$

12.8   From (12.8)

$$\frac{1}{N_e} = \left[\left(\frac{1}{1}\right)^2 + \left(\frac{1}{2}\right)^2 + \left(\frac{1}{4}\right)^2\right]^{1/2}$$

$$= \left[\frac{21}{16}\right]^{1/2} = 1.14$$

N_e = 0.88 cycles/mm.

12.9

$$N_e = 10 + \int_{10}^{20} (2 - 0.1f)^2 \, df$$

= 11 Hz, obviously worse than another system at Ne = 15 Hz.

12.10   Change of light would be $1 - 10^{(1.3 - 1.0)/2.0} = 41\%$

12.11   R/10 min = $(2 \times 10^{-7})(10)(10)(60)/[(0.5)(0.03)(0.25)^2(0.05)^2]$ = 512 R = 5.12 Gy

12.12   From (12.15), $d^2(C - 0.05)^2 = K$. Assume K = 1, then $d = (C - 0.05)^{-1}$ mm. Note the inverse relation between d and c for a fixed detection probability.

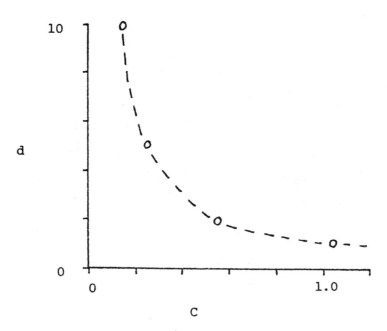

12.13 Exposure estimate is $= 2 \times 10^{-7}/(QDE)(RL)d^2(C - 0.05)^2$

Assume val. of QDE = 20%, $\quad 2 \times 10^{-7}/(20\%)(4\%)(0.5)^2(0.5 - 0.05)^2$
$\quad\quad$ 0.49 mR for d = 0.5 mm and C = 50%

$\quad\quad 2 \times 10^{-7}/(20\%)(4\%)(0.2)^2(0.10 - 0.05)^2$
$\quad\quad = 250$ mR for d = 0.2 mm and C = 10%

12.14 The number of levels to be digitized is not the maximal or average number of photons/pixel but the ratio of the standard deviation of the number to its mean.

12.15 The range of Hounsfield units is ±1000 and systems can resolve to within 2–4 H units, an amplitude resolution of better than 0.2%. To resolve to 1% or ±10 H units, the number of photons per voxel (volume pixel) must be greater than 10,000 as the images are obtained by processing many passes of the beam. Depending on the image reconstruction method, the number can range up to 10^6 per voxel. Because of tissue attenuation, the number incident to the surface of the patient must exceed that number times the number of pixels per scan line (resolution elements along a line through the patient), about 512, a total of almost 10^9 photons per surface element. Elements of any size require the same number of incident photons so that the number of photons/unit area increases if the voxels become smaller.

The eye can resolve 6 bits of dynamic range, close to 1%. The computer can integrate and expand data (window width and level) so that a greater total number of amplitude steps can be resolved but only 50 or so displayed in a single portion of the image.

12.16
$W_{14}, 0.1$; $W_{15}, 0.5$
$W_{24}, 0.4$; $W_{25}, 0.2$
$W_{33}, 0.1$; $W_{34}, 0.5$
$W_{43}, 0.4$; $W_{44}, 0.2$
$W_{52}, 0.1$; $W_{53}, 0.5$
$W_{62}, 0.4$; $W_{63}, 0.2$
$W_{72}, 0.6$

12.17 By measuring the output, the best assumption we can make is that all μ_{ij} are equal. Then from (12.16)

$$\ln\left(\frac{I_0}{I^{\theta k}}\right) = \sum w_{ij}^{\theta k} \mu_{ij}$$

$$\ln 2.0 = (4.2)\mu_{ij}$$

$\mu_{ij} = \dfrac{0.69}{4.2} = 0.16$ for all pixels intersected. Because no information is provided about pixels that are not intersected we cannot guess other μ_{ij}'s.

12.18 The sums of the projection data for x and y are equal, as shown

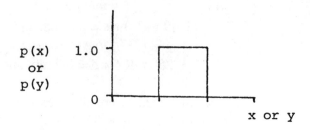

The sum of the projection data (1.0 or 0) is divided evenly over each block (1/3 or 0) for x and y and the results added to yield the backprojection shown.

0	1/3	0
1/3	2/3	1/3
0	1/3	0

12.19 $v = d\phi/dt$, $= 0.01$ m/s $\times 2.0 = 0.02$ V/m of blood vessel which should be harmless.

12.20 A logic gate driven by a flip-flop can remenber previous states and discriminate the three pulse heights shown.

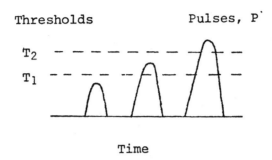

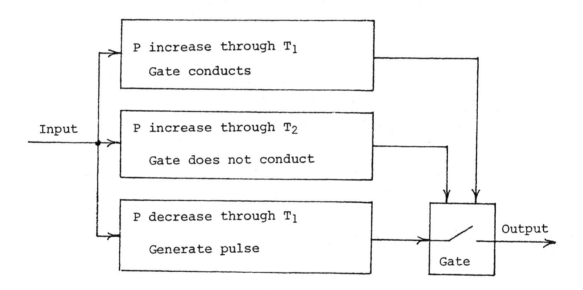

12.21 +y signal small
      −y signal large
      +x signal small
      −x signal large

12.22 From section 12.10 and Fig. 12.23,

Fourier transform in time $\quad F(\omega) = \int_{-\infty}^{+\infty} f(t) e^{-j\omega t} dt$

Fourier transform in space $\quad S(f) = \int_{-\infty}^{+\infty} L(x) e^{-j\omega x} dx$

Given $L(x) = k e^{-2|x|}$, which looks like fig. 12.28

To remove absolute value signs, break in two parts. For x negative, change sign of x to $-x$, so resulting quantity is positive.

$$S(f) = \int_{-\infty}^{0} k e^{(-2)(-x)} e^{(-j\omega x)} dx + \int_{0}^{+\infty} k e^{(-2)(+x)} e^{(-j\omega x)} dx$$

$$= k\left[\int_{-\infty}^{0} e^{(2-j\omega)x} dx + \int_{0}^{+\infty} e^{(-2-j\omega)x} dx \right]$$

From tables $\int e^{+ay} dy = \frac{e^{ay}}{a}$

$$S(f) = k\left[\frac{e^{(2-j\omega)x}}{2-j\omega} \Big|_{-\infty}^{0} + \frac{e^{(-2-j\omega)x}}{-2-j\omega} \Big|_{0}^{+\infty} \right]$$

$$= k\left[\frac{1}{2-j\omega} - 0 + 0 - \frac{1}{-2-j\omega} \right]$$

$$= k\left[\frac{1}{2-j\omega} + \frac{1}{2+j\omega} \right]$$

$$= k \frac{4}{4+\omega^2}$$

$$= k \frac{4}{4+2^2\pi^2 f^2} = \frac{k}{1+\pi^2 f^2}$$

f in cycles/cm

Medical Imaging Systems 119

12.23 From section 12.9, the amplifier must increase gain by the factor exp(kct), where t is reset to 0 at the beginning of each pulse.

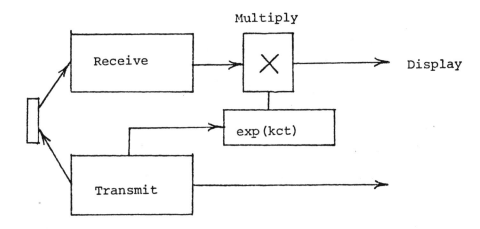

12.24 For Fig. 12.26, the round trip acoustic distance is about 0.02 m. Acoustic velocity of water is 1500 m/s.

Time, $t = d/v = 0.02/1500 = 13.3$ μs

Choose a sweep speed of 2 μs/cm or 20 μs across the 10 cm display.

Maximum repetition rate is 1.20 μs = 50 kHz.

12.25 The velocity of sound in bone is greater than in surrounding tissue so that the images of bones are foreshortened. The velocity of sound in air is less than the surrounding tissue but air is a less efficient conductor of the energy and the reflections of the surrounding tissue dominate the image. In fact, tissue images on the far side of the air cavities may not be truly visible, "blanked" by the presence of the air cavity.

12.26 To account for the inverse square relation of exposure vs. distance, s:
R/10 min = 5.12 Gy $(100/70)^2$ = 10.45 Gy, about double.

12.27 Solve for $\mu_{avg} = (\mu_1 + \mu_2)/2$, then find x for $I = I_0/2$ in eqn:
$I = I_0 e^{-\mu x}$, $\mu = [\ln (I_0/I)]/x$

$\mu_{avg} = 0.213$, HVL $= x = (\ln 2)/\mu_{ave} = 0.323$ mm Al,
Output = 380/60 = 6.33 mR/mAs

Chapter 13
Therapeutic and Prosthetic Devices

Michael R. Neuman

13.1   (a)   The energy in a single pulse will be

$$\frac{(5\text{ V})^2}{500\Omega} \times 2 \times 10^{-3}\text{s} = 10^{-4}\text{ W–s} \tag{P.13.1.1}$$

In five years time at a pulse rate of 70 pulses per minute, the total energy becomes

$$10^{-4}\text{ W-s} \times 70\text{ min}^{-1} \times 60\text{ min/hr} \times 24\text{ hr/da} \times 365.25\text{ da/yr}$$
$$\times 5\text{ yr} = 184 \times 10^4\text{ W–s} \tag{P.13.1.2}$$

(b)   If the pacemaker's efficiency is 35%, the energy supplied from the power source over a five year period must be

$$\frac{1.84 \times 10^4\text{ W–s}}{0.35} = 5.25 \times 10^4\text{ W–s} \tag{P.13.1.3}$$

(c)   This energy is supplied by two 2.8-V lithium cells. To determine the current from the battery, we divide this energy by the battery voltage

$$\frac{5.26 \times 10^4\text{ W·s}}{2 \times 2.8\text{ V}} = 9.39 \times 10^3\text{ A·s} \tag{P.13.1.4}$$

which gives the capacity in Ampere-seconds for the battery which can then be converted to Ampere-hours.

$$\frac{9.39 \times 10^3\text{ A·s}}{3600\text{ s/h}} = 2.61\text{ A·h} \tag{P.13.1.5}$$

(d)   There are several reasons why this pacemaker does not necessarily have its predicted ten-year life-time. The leakage current from the battery can increase due to external problems. Degradation of the packaging of the pacemaker when it is implanted can result in the development of current leakage paths around the battery and ruin the electronic circuit itself in older nonhermetically sealed pacemakers. This increased leakage current discharges the battery more rapidly than predicted, thereby shortening the overall life of the pacemaker. Most battery capacities are calculated with the battery operating at room temperature. When the temperature is elevated as it would be with the batteries implanted in the body, the battery capacity may be reduced, thus it is important to make sure in this problem that the batteries used in the pacemaker have the rated 2.61 A·h capacity at 37°C. It is important to note that the premature discharge of the batteries is much less of a problem today than it was several years ago when mercury batteries and nonhermetically packaged pacemakers were used.

13.2 While it is not possible to gain direct access to an implanted cardiac pacemaker, one can evaluate it's operation by several indirect means. These include the placement of electrodes on the patient's skin and analyzing the pacemaker pulse rather than the electrocardiogram. This should not be done with an electrocardiograph due to the fact that this instrument would be unable to respond to the high-frequency components of the pacemaker pulse. A more desirable device to use would be a high-frequency oscilloscope which can be synchronized to the pacemaker pulse. With this oscilloscope one can look at the pulse rate and the pulse duration. The pulse rate can give information about the condition of the pacemaker battery and, in some cases, the load seen by the pacemaker. In some cases the pulse duration is also affected by these circuit variables.

One can also evaluate the position of the stimulating electrodes by determining the vector for the pacemaker stimulus in the same way that it is done for the normal electrocardiogram. The vector will be a straight line parallel to a line joining the pacemaker electrodes. Thus, if a vector that has a direction that is different from that prescribed by the electrode position is found, it is likely that these are a result of electrode problems, either dislodgement of the electrodes or electrode or lead wire breakage.

13.3 The major limitation in the use of synchronous type cardiac pacemakers is due to electrical artifact picked up by the pacemaker and being misinterpreted by it as a heart beat. This can either disable the pacemaker, in the case of the demand pacemaker, or trigger the pacemaker as if an atrial beat had occurred, in the case of the atrial synchronous pacemaker. In either situation this can produce a serious arrhythmia rather than avoiding it. Thus it is important that the pacemaker be protected from all the external sources of interference. Radio-frequency pickup is one form of interference most frequently encountered, and so to avoid this, there must be adequate radio-frequency by-passing of all leads that could serve as antennae. The electronic circuit itself, especially the timing and synchronizing circuits, should be shielded in a metal container to avoid undesirable radio-frequency pickup.

13.4   The Figure below shows two key voltages associated with this demand pacemaker.

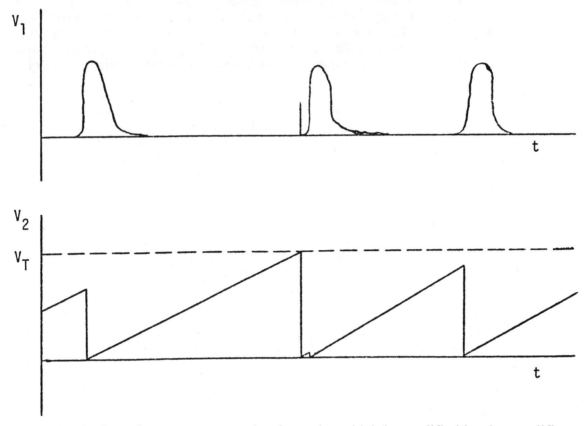

V_1 is the voltage seen across the electrodes which is amplified by the amplifier. These electrodes are placed in or on the ventricle, and the broad pulses shown represent normal ventricular depolarization. Note that the interval between the first broad pulse and the second one is longer than desired, and what has really happened is that the pacemaker detected this abnormally long R–R interval and caused a stimulus to be applied. This stimulus, the narrow pulse preceding the second broad pulse, in turn, caused the ventricles to contract producing the heartbeat immediately following it. The next beat, however, occurs at a normal range of R–R intervals, so the pacemaker does not stimulate the ventricles in this case.

The voltage V_2 in the Figure is the voltage at the reset circuit in the block diagram of Figure 13.3. Note that with each ventricular beat, or artificial stimulus, the reset circuit is reset and a voltage ramp is started at zero. This ramp increases linearly with time approaching a threshold voltage V_T. If the time is sufficiently great to allow the ramp to reach the threshold voltage, the reset circuit causes the timing circuit to produce a stimulus which is applied through the output circuit to the ventricles through the electrodes. Thus, the ventricles will be stimulated after the passage of time allowing the ramp voltage to reach the threshold voltage. If a naturally occurring beat is seen before that time, the ramp is resent and no stimulus is produced by the timing circuit.

13.5 The modification of the atrial synchronous pacemaker of Figure 13.4 involves incorporating features of the demand pacemaker shown in Figure 13.3. This is indicated on the block diagram reproduced below.

In addition to all of the blocks contained in the original design, three blocks have been added to supply stimulus pulses when appropriate atrial pulses are not produced. These blocks consist of a ramp generator which is reset by the occurrence of an atrial stimulus. The ramp generator is also connected to a threshold detector which, after a sufficient period of time when the ramp reaches the threshold, will trigger the 120 ms or 2 ms and the 500 ms multivibrators. The 500 ms multivibrator renders the pacemaker refractory to atrial stimuli for 500 ms. It is essential that this be in operation in the demand mode since ventricular stimulation can also trigger atrial contraction which could, in turn, cause the pacemaker to produce a second ventricular stimulus which might occur during the vulnerable period surrounding the T wave and produce ventricular fibrillation. The remainder of the operation of this circuit is identical to that in Figure 13.4.

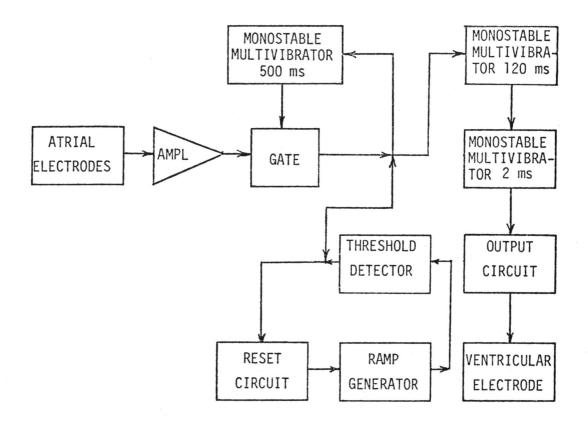

13.6 If the two stimulators have electrode systems which appear as identical load resistances, we can look at the power required to produce one minute's worth of pulses from each stimulator. Since the pacemaker in Problem 13.1 had a 500-Ω load resistance, let us use the same value for the bladder stimulator in this problem. Let us also require that both stimulators produce 5-V pulses. Calculating the energy required for a minute's worth of stimuli for the bladder stimulator, we get

$$\frac{(5 \text{ V})^2}{500 \text{ }\Omega} \times 2 \text{ ms} \times 100 \text{ sec}^{-1} \times 60 \text{ sec/min} = 0.6 \text{ W-s/min} .$$

(P 13.6.1)

The energy contained in a minute's worth of 5-V pacemaker pulses is

$$\frac{(5 \text{ V})^2}{500 \text{ }\Omega} \times 2 \text{ ms} \times 70 \text{ min}^{-1} = 0.007 \text{ W-s/min} .$$

(P 13.6.2)

Thus, it is seen in comparing the energy in the two cases that the bladder stimulator produces pulses with approximately 86 times the energy of the pacemaker pulses in one minute due to the increased rate. Since individual pulses for each stimulator are identical, this increased energy represents the increased pulse rate to the bladder stimulator. The actual energy required from the power supply is dependent on the efficiency of each stimulator. If we assume the same efficiency, it is clear that the bladder stimulator will require a much larger power supply capacity for the same lifetime as the cardiac pacemaker. For this reason, many bladder stimulators do not use primary battery power sources, but rather use secondary cells which can be recharged by transcutaneous radio-frequency techniques. In this way it is not necessary to use a battery which is 86 times larger.

13.7 From the definition of the mutual inductance of a transformer and the relationship between the coupling coefficient, the mutual inductance, and the self inductances of each coil, it can be shown that when other factors remain constant, the secondary voltage will vary in proportion to the coupling coefficient. Thus the voltage induced in the secondary of the transformer in Figure 13.5 will be given by

$$V_s = 10 \text{ V rms} \times 1.2 \times 0.5 = 6 \text{ V rms}$$

(P.13.7.1)

If this voltage is half-wave rectified by the Diode, it will produce a pulse having an amplitude of $6 \text{ V} \times \sqrt{2} = 8.49 \text{ V}$.

To consider the load impedance seen by the oscillator, we must take into account the fact that the Diode acts as a switch connecting the electrodes to the transformer only during the positive half cycle of the voltage appearing across the Diode. During the other half cycle, the Diode is an open circuit, and so if we assume the transformer to be semi-ideal (it cannot be ideal with a coupling coefficient of 0.5) the open circuit in the secondary will be reflected to the primary giving an infinite load impedance.

During the half cycle where the Diode is conducting, the 500 Ω load appears across the secondary, and this must be reflected to the primary. Since the transformer is not ideal due to a nonunity coupling coefficient, we cannot multiply this load resistance by the square of the turns ratio to get the impedance seen at the primary. Instead, we must use the relationship for reflecting a resistive load to the primary of a nonideal transformer

$$Z = \frac{\omega^2(M^2 - L_1L_2) + j\omega R_L L_1}{R_L + j\omega L_2}$$

(P 13.7.2)

Since

$$M = k\sqrt{L_1L_2}$$

(P 13.7.3)

When we substitute this in Equation (P.13.7.2) we get

$$Z = \frac{\omega^2 L_1 L_2 (k^2 - 1) + j\omega L_1 R_L}{R_L + j\omega L_2}$$

(P 13.7.4)

If we take the liberty of assuming our transformer to be quasi-ideal, namely that the self inductances of each coil are large with respect to the load resistance, R_L, this equation, when we substitute the value for k, can be simplified to

$$Z \approx \frac{L_1}{L_2} R_L + j\, 0.75\, \omega L_1$$

(P 13.7.5)

For an ideal transformer, the self-inductance ratio is the same as the square of the turns ratio, this equation can be rewritten as

$$Z \approx \left(\frac{N_1}{N_2}\right)^2 R_L + j\, 0.75\, \omega L_1$$

(P 13.7.6)

13.8 Most implantable stimulators for the suppression of pain as well as for other electrical stimulation purposes, are the constant voltage type of stimulator. this means that the voltage amplitude of stimulus pulses is constant during the pulse and the current from each pulse will be a function of the effective load. If the load resistance increases with time, we will see a similar decrease in electrode current with time. This means that the total charge transfer at the electrode will diminish with time and since it is really the charge which is transferred that determines whether a stimulus is effective or not, as the amount of charge transfer drops, the stimulator may begin to no longer stimulate the excitable tissue.

This problem is best minimized by utilizing a stimulator which produces constant current rather than constant voltage stimulus pulses. In this case, the amount of charge transferred at the electrodes will be unaffected by electrode impedance changes. An increase in electrode impedance will only increase the pulse voltage. It is not always practical to produce implantable constant current stimulators since they require higher voltage power sources than do the constant voltage stimulators. Thus, for a practical stimulator circuit, a compromise between the two types of stimulators must be reached to maintain efficacy. This compromise involves producing a relatively high-voltage pulse with a high source impedance to have the charge transferred during a stimulus pulse be less effective by electrode impedance than it would be for the constant voltage stimulator.

13.9 The discrepancy is primarily due to losses in the circuit between the capacitor and the patient. These losses result from resistances found at (1) the electrode-patient's skin interface, (2) resistance in the series inductor for increasing the discharge duration, and (3) resistance in the lead wires.

The best way to minimize the discrepancy is to minimize each of these resistances. The electrode-skin interface resistance can be minimized by using large surface area electrodes which make good contact with the skin and the contact is augmented with an electrolyte gel. Resistance, and core losses for that matter, in the inductor can be reduced by winding the inductor on a large core which will not saturate at the maximum current using large-diameter (low-gage) copper wire. Similarly, the lead wire resistance can be minimized by using lead wires with large effective cross-sectional area.

13.10 From Equation 13.4 we see that the energy stored in the capacitor of a defibrillator is proportional to the square of the voltage. Thus, if we were to, say, double the voltage of a defibrillator, we would only need a capacitor that was 1/4 as large to store the same energy. Since the size of a capacitor usually is approximately proportional to its capacitance, and to its maximum voltage, it is possible by operating the defibrillator at higher voltages to get by with a capacitor which is physically smaller. When you do this, nevertheless, you must be concerned about the size of the power supply to obtain the higher voltage. You may find that nothing is gained in terms of size and mass when that consideration is made. Operating the defibrillator at higher voltage and, hence, smaller time constants, can also maximize the peak current of the discharge. This is due to the relative constant magnitude of the thoracic resistance which limits the current. At higher voltages, the limiting value of the current will be higher. Higher voltage pulses from the defibrillator can put the operator at additional risk. Higher voltage pulses require better insulation on the paddles to protect the person applying the pulse.

13.11 If the electrodes are not in good contact with the skin, the effective cross-sectional area of the electrode may be reduced. This means that the current density over part of the skin under the electrode will be much higher than over the rest of the skin under the electrode. The effective resistance of the electrode will also increase. All of these effects will add up to produce greater heating of that region of the skin where the current is flowing. Currents may actually be high enough to produce burning of the skin. If the electrodes are making very poor contact with the skin, the effective load resistance on the defibrillator may be sufficiently high so that relatively small currents flow which will not produce effective defibrillation. This high resistance essentially lengthens the time constant of the capacitive discharge, resulting in a much longer discharge waveform. If the contact between the electrodes and the skin is really poor, there might even be a high-voltage discharge across the gap between the electrodes and the skin dissipating most of the energy in the spark rather than in the patient. The spark can also produce intense local burning of the patient's skin. In this case, the waveform of the discharge will be considerably distorted and irregular due to the high-voltage discharge. Finally, if the defibrillator is discharged before the electrodes are making contact with the skin, the full capacitor voltage will appear across the electrodes. As the electrodes are brought close to the skin, a spark will be produced before the electrodes contact the skin, which can produce burning as well as dissipating some of the energy intended for the patient. It is for this reason that most defibrillator electrodes are arranged so that they are only connected when they are firmly pressed against the skin.

13.12 When the patient to be defibrillated has an operating cardiac pacemaker implanted, there are two principal problems which might be encountered during defibrillation. One is that the high-energy defibrillation pulse might damage the electronic circuits of the pacemaker. This is especially true in cases of synchronous pacemakers where the pacemaker is designed to sense relatively low cardiac voltages and to act upon them. The high voltage that can sometimes result from the defibrillator discharge could damage the sensing circuit in this pacemaker. The second problem that can occur is that a pacemaker pulse might occur at the critical time following defibrillation and cause the patient to fibrillate again. In the case of the synchronous pacemaker, the defibrillation pulse might trigger the pacemaker and produce a ventricular stimulus immediately following the defibrillation pulse before the ventricular cells have had a chance to completely repolarize. This, too, could result in reestablishing ventricular fibrillation.

In most cases there really isn't much that can be done other than to be aware of these problems in patients with implantable pacemakers who have undergone cardiac arrest. Although synchronizing the defibrillator to the pacemaker can avoid these problems, unless a cardioverter that is capable of being triggered by the pacemaker stimulus is immediately available, this mode of action might put the patient at more risk than would the reestablishment of ventricular fibrillation because of the time delay necessary to set up the cardioverter and synchronize it to the pacemaker artifact. In cases where the pacemaker might be permanently damaged, this is not such a serious risk. A temporary transvenous pacemaker can be placed on the patient once his heart has been defibrillated and the defective cardiac pacemaker removed and a new one implanted within the next few days once the patient's vital signs have stabilized. It is better to have a burnt-out pacemaker than a burnt-out patient.

13.13 (a) When the defibrillator is attached to the patient, we have a series RLC circuit with an initial charge on the capacitor. The differential equation governing this circuit is

$$L\frac{di}{dt} + R_L i + \frac{1}{C}\int i\, dt + V_c = 0$$

(P 13.13.1)

Basic circuit theory tells us that the solution for this equation will give a critically damped current when

$$\frac{R_L^2}{4L^2} = \frac{1}{LC} \tag{P 13.13.2}$$

Thus, for these conditions to be met, the inductance must be given by

$$L = \frac{R_L^2 C}{4} = \frac{(50\Omega)^2 (2 \times 10^{-5} \text{ F})}{4} = 12.5 \text{ mH} \tag{P 13.13.3}$$

(b) Again, from basic circuit theory, we can show that for the critically damped case the current is given by

$$i = \frac{V_c}{L} t \, e^{-\frac{1}{\sqrt{LC}} t} \tag{P 13.13.4}$$

To determine the maximal current, we take the time derivative of this current

$$\frac{di}{dt} = \frac{V_c}{L}\left[e^{-\left(\frac{1}{\sqrt{LC}}t\right)} - \left(\frac{1}{\sqrt{LC}}t\right) e^{-\left(\frac{1}{\sqrt{LC}}t\right)}\right] = 0 \tag{P 13.13.5}$$

The time at which the maximum occurs is

$$t = \sqrt{LC} \tag{P 13.13.6}$$

Thus, the maximal current will be

$$i_{max} = \frac{V_c \sqrt{C}}{\sqrt{L}} e^{-1} \tag{P 13.13.7}$$

To determine the initial voltage on the capacitor, we must apply Equation 13.4

$$200 \text{ J} = \frac{1}{2} (2 \times 10^{-5} \text{F}) V_c^2 \tag{P 13.13.8}$$

$$V_c = 4472 \text{ V} \tag{P 13.13.9}$$

Now, when we substitute this into the equation for the maximum current, we get

$$i_{max} = \frac{4472 \text{ V} (2 \times 10^{-5} \text{ F})^{1/2}}{(1.25 \times 10^{-2})^{1/2} \, e} = 65.8 \text{ A} \tag{P 13.13.10}$$

13.14

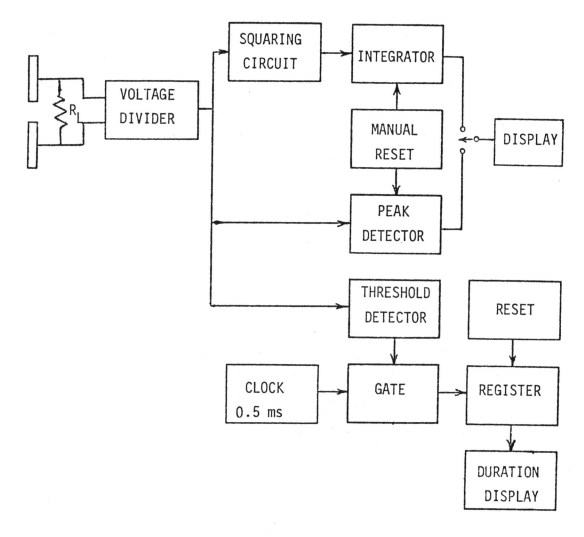

A defibrillator tester which measures the energy of a discharge, the peak voltage of the discharge, and the duration of the discharge, is shown in block diagram form. This instrument consists of two metal plates which are connected to a high-power load resistance R_L. This resistance must be capable of dissipating the energy of the defibrillator pulse and should not have any reactance associated with it. In operation, the defibrillator paddles are placed against these metal plates, and the defibrillator is discharged through R_L. A voltage-divider circuit connected across this load resistance provides a low-voltage output which is proportional to the voltage seen across the load. This voltage is squared by the squaring circuit and then integrated over the duration of the pulse. The resulting integral is directly proportional to the energy in the pulse, and the output of the integrator is displayed on an analog meter. The energy scale on this meter will be linear since the voltage was fed directly to the integrator without squaring it, one could still display the energy of the pulse on the meter, but it would be necessary to have a nonlinear meter scale. A manual reset establishes zero initial conditions for the integrator to prepare it for the next defibrillation pulse.

The second part of the instrument determines the peak voltage of the pulse by connecting the attenuated pulse waveform to a peak detector. The output of the peak detector can also be shown on the analog display device. It, too, must be reset to zero before each pulse is measured.

The attenuated voltage is also fed to a threshold detector having a threshold voltage slightly above zero. The threshold detector goes positive over the duration of the pulse and enables a register to count 0.5 ms clock pulses. Following the end of the defibrillation pulse, the number in the register corresponds to the duration of the defibrillation pulse measured to the nearest 0.5 ms. This is indicated on the digital duration display. A manual reset is used to reset the register for measurement of the next pulse.

13.15 The essence of this design is to draw the gas out of the aortic balloon during systole or just prior to it and to then fill the balloon with gas during diastole making sure that the balloon is completely filled before the next systole occurs. Thus an important aspect of the system is the signal upon which the balloon pumping device triggers. Three signals offer possibilities for such triggering: the QRS complex of the electrocardiogram, the first heart sound, and the rising phase of the arterial blood pressure waveform. Each of these has advantages and disadvantages.

The electrocardiogram is the easiest of all the signals to obtain, and the QRS complex occurs just prior to ventricular contraction. Thus, by detecting the QRS complex and delaying the response to the detection of this event for a few millisecond to allow the ventricle to begin to contract, appears to be a reasonable way to trigger the suction valve on the balloon controller. Indeed this is true, but one must always worry about artifact on the electrocardiogram and whether this can produce false triggering which might be detrimental.

A possible system for doing this is illustrated in the block diagram given below. The electrocardiogram, after being appropriately amplified, is differentiated and passed through an absolute value circuit. The maximum rate of change of the electrocardiogram occurs during the QRS complex and so the absolute value circuit will have its maximum output voltage during this time. A self adjusting peak detector circuit can then detect the actual QRS complex. The self adjusting peak detector is used rather than a conventional peak detector or a threshold detector so that as the amplitude and morphology of the electrocardiogram changes, the peak detector threshold will also change. A convenient value to adjust the peak detector threshold is 75% of the peak voltage of the previous QRS complex. In this way, variations in QRS complex amplitude will still be detected while maintaining the highest possible threshold to minimize accidental detection of artifact. The output of the peak detector triggers monostable multivibrator #1 which provides several milliseconds of delay before monostable multivibrator #2 is triggered. This multivibrator has a pulse duration corresponding to the duration of systole, and it also drives the suction valve to the balloon. When the output of this multivibrator returns to its low value, an inverter and driver open the pressure valve to inflate the balloon.

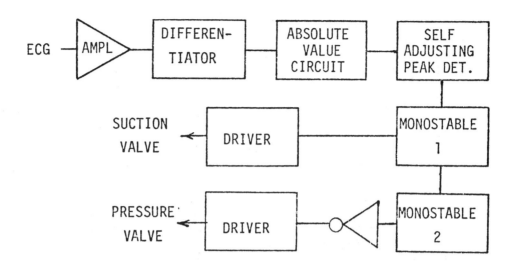

The second signal which can be used is the first heart sound. This occurs at the beginning of systole when the mitral and tricuspid valves close due to the increase in ventricular pressure. This is a very appropriate time for suction to be applied to the balloon since this will reduce the aortic pressure and cause earlier opening of the aortic valve than would normally occur without the balloon. If the balloon is collapsed earlier than this, the reduction in aortic pressure during diastole would mean that there would be diminished perfusion of the coronary arteries. Since one of the reasons for cardiac assist devices of the counter-pulsation type is to increase coronary perfusion, using the first heart sound to trigger the balloon collapse seems quite attractive. The second heart sound could then be used to trigger the reinflation of the balloon. A problem with this system, which is even more evident than it is with the electrocardiogram control, is the presence of artifact. In this case extraneous sounds can cause triggering of either the suction or filling phases of the balloon's operation and may even cause the balloon to oppose the activities of the heart rather than to assist them. For example, if a single extraneous sound is accidentally recognized as the first heart sound, when the true first heart sound comes, it may be recognized as the second heart sound and cause the balloon to be filled rather than emptied. This would greatly increase the aortic pressure at the base of the heart and cause the myocardium to work all the harder, thereby giving the opposite result from that desired.

The third signal which can be used to trigger the balloon is the arterial pressure waveform itself. If one uses the rising phase of the arterial pressure to trigger the deflation of the balloon and then waits until the dicrotic notch occurs in the arterial pressure waveform to reinflate the balloon, one can assist the heart. The problem is that if we wait until the rising phase of the arterial pressure waveform occurs before we deflate the balloon, we will have not decreased the opening pressure of the aortic valve, since the valve must be open for the arterial pressure increase to occur. Thus, the isovolumnic phase of myocardial contraction will remain unchanged. It is more desirable to reduce the aortic pressure just prior to opening the aortic valve so that it will be open at a lower pressure thereby reducing the ejection pressure at the left ventricle. Even if this is not done, the increased pressure during diastole will help to increase coronary perfusion.

132 Therapeutic and Prosthetic Devices

13.16 The pump oxygenator system can be monitored with various types of instrumentation. The specific instrumentation and the control system is dependent on each individual type of pump oxygenator. Nevertheless, a general system is illustrated below.

The principal function of the pump is to maintain cardiac output and systemic blood pressure. By sensing the pressure just beyond the pump output and displaying this pressure as well as using it to control the pump, objective data from manual or automatic control is obtained. The instrumentation system involves a strain gage pressure sensor placed at the output of the pump which is amplified and passed through a low-pass filter having a corner frequency of less than the pump pulse rate for a pulsatile pump to give the mean pressure. This pressure is displayed and used to control the pump through a servo controller. A similar pressure sensor is also coupled to the central venous circulation through a fluid-filled catheter and its output is amplified and displayed. In the case of the central venous pressure, this signal is not used to control the pump, nevertheless, abnormally high central venous

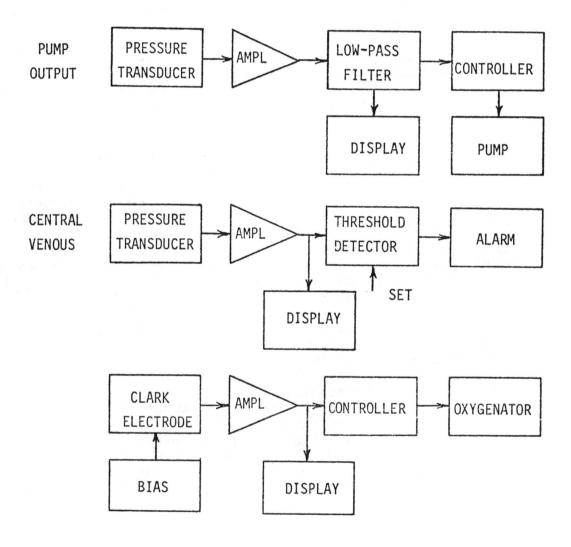

pressures represent a danger to the patient and so the output of the central venous pressure measurement system is fed to a threshold detector which is set to indicate a maximum pressure at which point an alarm is sounded.

Blood oxygenation is sensed just beyond the oxygenator by a Clark electrode. This measures the partial pressure of oxygen in the blood, and the signal is amplified and displayed along with the other variables. This signal can also be used to control the oxygenator through a second servo control loop. In this case, variables such as gas entering the oxygenator or effective cross-sectional area of the oxygenator can be controlled by this controller.

It is important to point out that all of the sensors described above are subject to a certain amount of drift and instability and to close the loop with a servo controller, as pointed out, might put the patient at more risk than if the loop were left open. This is due to the possibility of artifact or drifting signals causing the controller to operate the pump or the oxygenator in a detrimental mode. This it is important in any design that adequate safeguard features are included in any automatic control system so that they offer minimal risk to the patient.

13.17 The instrumentation system to provide the required protection for the artificial kidney system is relatively straightforward. It really consists of two alarm detectors: one on the blood system and one on the dialysate system. In the case of the blood alarm, this will detect major leaks in the dialyzer membrane. Since the blood is maintained at a higher pressure than the dialysate, should a major leak occur, this would cause a significant drop in the blood pressure. Thus a pressure sensor is connected to the blood line to sense the blood pressure, and its output is amplified and coupled to a threshold detector. When the pressure drops beneath a preset threshold, it sends an enable signal to the alarm circuit and sets off the alarm.

Most leaks in such an artificial kidney system, however, will not be large enough to produce a significant drop in blood pressure to be detected by this type of a safety circuit. Instead, it is possible to sense leaks by detecting blood in the dialysate itself. Dialysate is normally a clear liquid and so the presence of blood will give it a red color. In the block diagram shown, the red color is detected by beaming white light into the dialysate outlet from the dialysis chamber and looking at the reflected light. A red filter is used so that only reflected red light is considered. A photodetector sensitive to the red light is then used to drive an amplifier and a threshold detector. In this case the threshold detector responds to signals above a set point which correspond to reflection of red light from blood in the dialysate. This threshold detector enables the alarm. The types of photodetectors that can be used consist of various sensors described in Chapter 2. In the design, a semiconductor photodiode or phototransistor is recommended since it is small and has a moderate sensitivity which would be adequate in this application. It also will easily interface with other solid state electronic components in the amplifier and threshold detector.

134 Therapeutic and Prosthetic Devices

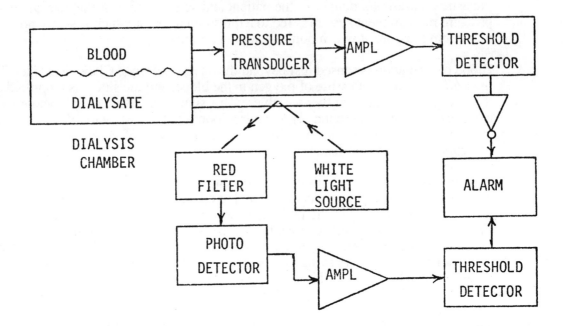

13.18 The modification of the system in Figure 13.15 consists of taking the outputs of the three scaling amplifiers to the left of the dashed line and connecting each to its own

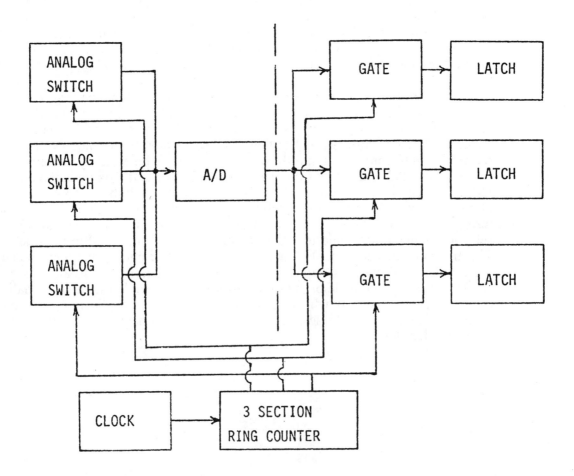

analog switch. The outputs from each switch are connected together and to the input of the analog-to-digital converter. Each switch is controlled by different outputs of a three-section astable ring counter. The counter is arranged so that only one output is high at a time, thus only one analog switch will be turned on at a time. A corresponding gate on the digital side of the analog-to-digital converter is turned on for each analog switch allowing the digital signal to be stored in a latch register. The output of this register is connected to the digital display and digital comparator and alarm as shown in Figure 13.15.

This system is a time-division multiplexing system, and you will note its similarity to the system described for multichannel radio telemetry in Chapter 6.

13.19 From Handbook of Chemistry and Physics, most plastics range from $\varepsilon = 2$ to 4. Choose $\varepsilon = 3$.

$$Z = \frac{1}{\omega C} = \frac{D}{2\pi f \varepsilon A} = \frac{0.002 \text{ in } (0.0254 \text{ m/in})}{2\pi \cdot 500000 \cdot 3 \cdot 8.8 \cdot 10^{-12} \cdot 0.007 \text{ m}^2} = 87 \, \Omega$$

13.20 A simple means of determining if there is a poor contact between the indifferent electrode of an electrosurgical apparatus and the patient, is to apply an additional or several additional skin electrodes on the patient. These electrodes should be at the same potential as the indifferent electrode of the electrosurgical apparatus at the band of frequencies produced by the generator. If they are not, this indicates that a high-impedance connection between the indifferent electrode and the skin exists and when current flows, a significant voltage drop is seen across this impedance. This voltage should be detected and used to disable the generator so that burns will not be produced.

Chapter 14
Electrical Safety

Walter H. Olson

14.1 The minimal current for ventricular fibrillation by macroshock is about 75 mA as shown in Fig. 14.1. For 10 cm × 10 cm cross-sectional area, the current density would be

$$\frac{75\text{ mA}}{100\text{ cm}^2} \times \frac{10^{-3}\,\mu\text{A}}{1\text{ mA}} \times 7.5\,\mu\text{A/mm}^2$$

The lowest current density on the graph in problem 14.8 is 12 μA/mm^2 for a 90 mm^2 catheter. These values are comparable. This comparison supports the view that macroshock and microshock cause fibrillation by the same mechanism.

14.2 The variation in rms current as a function of frequency for a constant applied voltage of 20 mV is given by

$$I_{rms} = \frac{V_{o\text{-}p}}{\sqrt{2}}\frac{1}{Z} = \frac{V_{o\text{-}p}}{\sqrt{2}}\sqrt{\frac{1+\omega^2 C^2 R^2}{R}}$$

f	2	10	100	500	700	1000	1500	2000	3000	5000	7000	8000	10,000
I$_{rms}$ mA	14.1	14.1	14.6	14.8	15.4	16.7	19.5	22.7	30.1	46.7	63.8	72.5	90.0

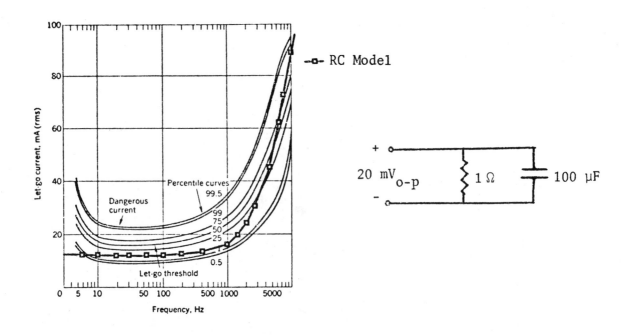

14.3 (a) Premature ventricular excitation (QRS waves) will occur in addition to atrial excitation at the time when the P wave would normally occur alone.

(b) Ventricular excitation (QRS waves) would occur at about the normal time although the QRS complex duration may be increased and atrial excitation may occur if atrial repolarization has occurred.

(c) Ventricular fibrillation may occur because this is the vulnerable period. If the shock is very intense, complete myocardial contraction may result. In this case a normal rhythm may return.

(d) Premature excitation of the atria and ventricles is likely followed by a normal rhythm.

Figure 14.4 shows that fibrillation threshold current increases for shock duration less than 1 second. Heart rate is also about 60 bpm or once per second. Thus shocks with durations less than about 1 s may not be present during the most "vulnerable period" (T wave) of the cardiac cycle.

14.4 Current flow for a conductive catheter. Note the current density does not become high at any point along the conductive catheter. In Fig. 14.5b all the current must flow out the end of the nonconducting catheter.

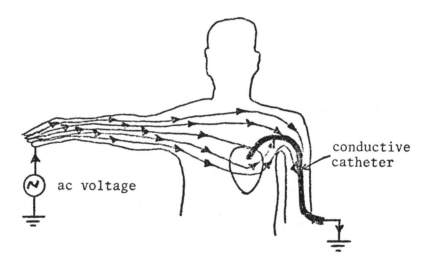

14.5 Removal of the ground from the water pipe at the transformer secondary center tap in Fig. 14.6 would at first glance apparently convert the 115-V circuits to isolated circuits such as the one shown in Fig. 14.7, however, the ground wires in Fig. 14.6 that are normally connected to equipment cases would not be grounded as they are in Fig. 14.7. Instead these equipment cases would be connected to the floating center tap on the transformer. A silent fault could occur from 115 V to ground. Then the equipment cases would be at 115 V above ground and undoubtedly some grounded objects such as water pipes would be available. A macroshock would result if someone touched both a floating equipment case and some object that was really grounded.

A primary-to-secondary fault in the main transformer (Fig. 14.6) could apply up to 4800 volts to patient equipment. Numerous equipment faults and shock hazards would probably result.

14.6 Impedance of the 0.3 µF capacitor is $X = 1/(j\omega C) = 1/(j2\pi 60 \cdot 0.3 \text{ µF}) = -j8.8 \text{ k}\Omega$. The resistor $R = 10 \text{ k}\Omega$. Voltages across these two components add at ≈90° so the source voltage driving the detector is ≈ 120 V(0.707) = 85 V. The impedance of the detector should be $Z = V/I = 85 \text{ V}/25 \text{ µA} = 3.5 \text{ M}\Omega$. So either change 33 kΩ to 3.5 MΩ or change 0.01 µF to $C = 1/(j\omega X) = 1/(2\pi 60 \cdot 3.5 \text{ M}\Omega) = 760 \text{ pF}$.

14.7 If either line faults to ground, current will flow for the time the switch is on the other line and be detected.

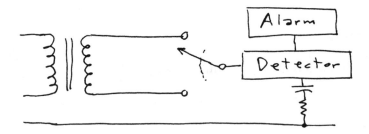

14.8 Hypothesis: Cell membrane current density raises resting potential to exceed threshold. Fig. 14.10 is replotted as 60-Hz fibrillation threshold current density versus catheter area.

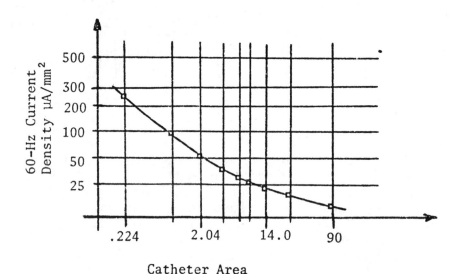

This replot does not support the hypothesis. Fibrillation current density is not independent of catheter area (curve above is not a horizontal straight line). CAUTION: We have implicitly assumed here that catheter current = cell membrane current. This may not be true due to complex geometries and tissue anisotropy. For example, there may be a membrane layer between the catheter tip and the excitable tissue. The hypothesis may be true for large areas where the current density in the excitable tissue is approximately equal to the current density in the catheter.

14.9 This is a microshock hazard so the maximal safe current is 10 µA. This equivalent circuit is:

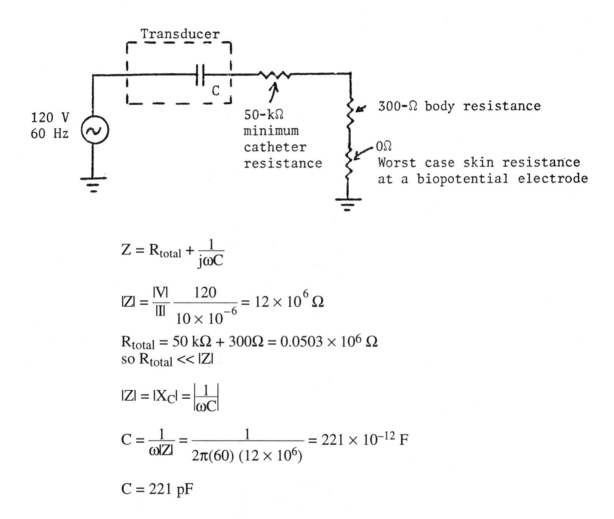

$$Z = R_{total} + \frac{1}{j\omega C}$$

$$|Z| = \frac{|V|}{|I|} \frac{120}{10 \times 10^{-6}} = 12 \times 10^6 \, \Omega$$

$R_{total} = 50 \text{ k}\Omega + 300 \Omega = 0.0503 \times 10^6 \, \Omega$
so $R_{total} \ll |Z|$

$$|Z| = |X_C| = \left|\frac{1}{\omega C}\right|$$

$$C = \frac{1}{\omega |Z|} = \frac{1}{2\pi(60)(12 \times 10^6)} = 221 \times 10^{-12} \text{ F}$$

$C = 221 \text{ pF}$

14.10 The catheter tip area = πr^2 = 4.01 mm². For this area Fig. 14.10 shows that the lowest current for fibrillation or pump failure was about 50 µA. (Note these data are for dogs and may not be reasonable for humans). Assume the catheter is purely resistive and cylindrical.

140 Electrical Safety

$$R = \frac{V}{I} = \frac{120}{50 \times 10^{-6}} = 2.4 \times 10^6 \, \Omega \text{ (neglect 300-}\Omega\text{ body R)}$$

$$R = \frac{\rho L}{A} \text{ where } \begin{array}{l} \rho = \text{specific resistivity} \\ L = \text{length} \\ A = \text{cross-sectional area} \end{array}$$

$$\rho = \frac{RA}{L} = \frac{(2.4 \times 10^6)(4.01 \text{ mm}^2)}{10^3 \text{ mm}} \times \frac{1 \text{ cm}}{10 \text{ mm}} = 962.4 \, \Omega \cdot \text{cm}$$

Saline resistivity is about 150 $\Omega \cdot$cm.

14.11

$$Z = 1 \, M\Omega + 300\Omega + 50 \, K\Omega + \frac{\dfrac{20 \times 10^6}{j120\pi \times 10^{-10}}}{20 \times 10^6 + \dfrac{1}{j120\pi \times 10^{-10}}}$$

$$|Z| = 1.05 \times 10^6 + \frac{\sqrt{R^2 + \omega^2 C^2 R^2}}{1 + \omega^2 C^2 R^2}$$

$$= 1.05 \times 10^6 + 15.9 \times 10^6$$

$$|Z| = 16.9 \times 10^6$$

$$I = \frac{V}{|Z|} = \frac{120}{16.9 \times 10^6} = 7.1 \, \mu A$$

Safe, since the safety limit is 10 μA.

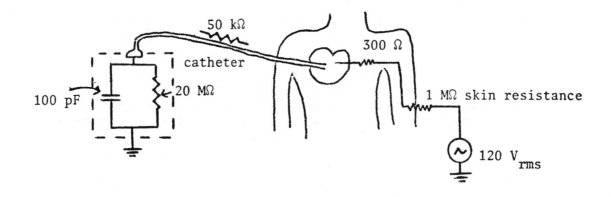

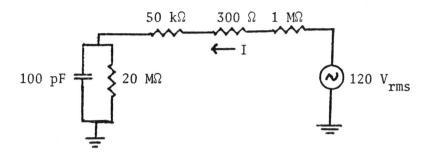

14.12 This is an open-ended problem so many solutions are possible. An example is a faulty intracardiac ECG monitor with a grounded metal case that can be touched by the patient.

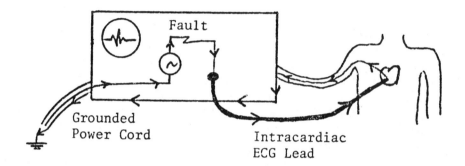

Assume the patient's skin resistance was high (1 MΩ)

$$I = \frac{120 \text{ V}}{1 \text{ M}\Omega} = 120 \text{ }\mu\text{A} \quad \text{Hazardous!}$$

14.13 This is an open-ended problem with many correct solutions most of which are variations of Examples 14.1–14.3. Each scenario must have (a) a source current, (b) a direct electric connection to the heart, (c) a return path for the current anywhere on or inside the body possibly including the heart, (d) pictorial diagram, (e) equivalent circuit, (f) appropriate safety tests Fig. 14.17–14.21. A common student error is to specify a device such as a metal surgical instrument for the electric connection to the heart. Often such devices would have conductive surfaces that would contact other body tissue so that a true microshock would not result.

142   Electrical Safety

14.14   12 mA from hot B to ground should trip. R = V/I = 120 V/12 mA = 10 kΩ.

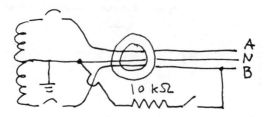

14.15   Eliminate the B winding and take power between N and A.

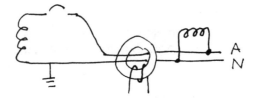

14.16   This open-ended design problem has no unique answer. As stated in the text an outlet can have 64 possible states. States that connect hot to neutral and/or ground can probably be ignored because the circuit breaker would trip. Some current limiting requirements should be incorporated into the design to avoid causing hazards at other locations such as the one depicted in Fig. 13.11. This would apply to detection of high resistance ground or neutral contacts. A load could be applied between hot and neutral to determine if one of these contacts is open. An independent reference ground connection is apparently needed. A block diagram for a hypothetical outlet tester (courtesy of W. S. Friauf, DEIB, DRS, NIH) is given below. The Emergency Care Research Institute, Plymouth Meeting, PA, has developed a high-quality outlet tester but details are not available at this writing.

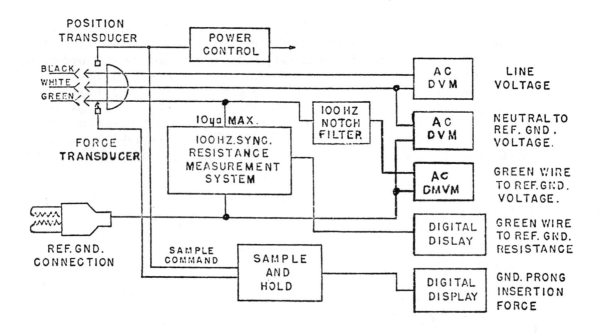

14.17

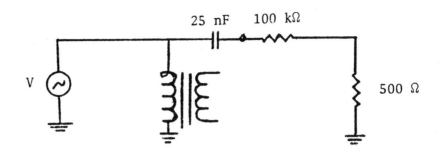

$$V = I|Z| = (75 \text{ mA}) \sqrt{R^2 + \frac{1}{\omega^2 C^2}}$$

$V = 10.9 \text{ kV}_{rms}$.

14.18 R_B supplies 15 V/1.5 MΩ = 10 μA all the time. R_G supplies ±10 V/1 MΩ = ±10 μA so reverse current through CR_3 is 0 to +20 μA with single polarity. Reverse current through CR_2 = 0 to 20 μA with a single polarity. 10 μA is supplied by R_B and ±10 μA by R_K, yielding $v_o = ±10$ V.

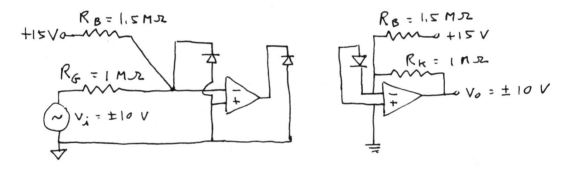

It is also possible to use two light-separated enclosures with light on top separated from light on bottom but that uses more parts.

Suggested Laboratory Experiments

John G. Webster

1. Measure nerve velocity from elbow to hand.
2. Record electroencephalogram and demonstrate alpha waves.
3. Record a 12-lead electrocardiogram.
4. Identify arrhythmias from prerecorded tapes.
5. Measure motion artifact from electrodes and from skin.
6. Construct an ECG amplifier from components.
7. Record a vectorcardiogram.
8. Transmit an ECG over telephone lines.
9. Transmit an ECG by radiotelemetry and measure range.
10. Measure blood pressure using a cuff.
11. Determine catheter response using the "pop" test.
12. Measure heart sounds and blood vessel wave velocity.
13. Measure vessel saline flow electromagnetically.
14. Measure blood flow ultrasonically.
15. Measure saline flow by a thermodilution catheter.
16. Measure volume change by impedance plethysmography.
17. Measure gas flow by a pneumotachograph.
18. Measure lung volumes using a spirometer.
19. Measure metabolic efficiency by oxygen consumption.
20. Measure lung volume by nitrogen washout.
21. Measure lung volume by helium dilution.
22. Measure gas transport by carbon monoxide diffusion.
23. Measure ratio of two chemicals with a spectrophotometer.
24. Determine buffer curve with pH meter.
25. Construct oxygen electrode and amplifier from components.
26. Demonstrate operation of a demand pacemaker.
27. Calibrate a defibrillator output.
28. Measure finger-to-thumb current perception vs. frequency.
29. Determine protection provided by isolation transformer.
30. Run NFPA tests on isolated and nonisolated equipment.
31. Determine oxygenation using a pulse oximeter.
32. Determine apnea using an apnea monitor.